Developing
# Essential Understanding
*of*
# Multiplication and Division
*for* Teaching Mathematics *in*
# Grades 3–5

Albert Dean Otto
Illinois State University
Normal, Illinois

Janet H. Caldwell
Rowan University
Glassboro, New Jersey

Edward C. Rathmell
*Volume Editor*
University of Northern Iowa
Cedar Falls, Iowa

Cheryl Ann Lubinski
Illinois State University
Normal, Illinois

Rose Mary Zbiek
*Series Editor*
The Pennsylvania State University
University Park, Pennsylvania

Sarah Wallus Hancock
Maplewood Richmond Heights
School District
Saint Louis, Missouri

NATIONAL COUNCIL OF
TEACHERS OF MATHEMATICS

Copyright © 2011 by
The National Council of Teachers of Mathematics, Inc.
1906 Association Drive, Reston, VA 20191-1502
(703) 620-9840; (800) 235-7566; www.nctm.org
All rights reserved

Third printing 2017

Library of Congress Cataloging-in-Publication Data

Developing essential understanding of multiplication and division for
teaching mathematics in grades 3-5 / Albert Dean Otto ... [et al.].
    p. cm.
  Includes bibliographical references.
  ISBN 978-0-87353-667-7
  1. Multiplication—Study and teaching (Elementary) 2. Division—Study
and teaching (Elementary) I. Otto, Albert Dean. II. National Council of
Teachers of Mathematics.
  QA115.D48 2011
  372.7'049--dc22
                                    2010053665

The National Council of Teachers of Mathematics supports and advocates for the
highest-quality mathematics teaching and learning for each and every student.

Printed in the United States of America

# Contents

# Foreword

Teaching mathematics in prekindergarten–grade 12 requires a special understanding of mathematics. Effective teachers of mathematics think about and beyond the content that they teach, seeking explanations and making connections to other topics, both inside and outside mathematics. Students meet curriculum and achievement expectations when they work with teachers who know what mathematics is important for each topic that they teach.

The National Council of Teachers of Mathematics (NCTM) presents the Essential Understanding Series in tandem with a call to focus the school mathematics curriculum in the spirit of *Curriculum Focal Points for Prekindergarten through Grade 8 Mathematics: A Quest for Coherence*, published in 2006, and *Focus in High School Mathematics: Reasoning and Sense Making,* released in 2009. The Essential Understanding books are a resource for individual teachers and groups of colleagues interested in engaging in mathematical thinking to enrich and extend their own knowledge of particular mathematics topics in ways that benefit their work with students. The topic of each book is an area of mathematics that is difficult for students to learn, challenging to teach, and critical for students' success as learners and in their future lives and careers.

Drawing on their experiences as teachers, researchers, and mathematicians, the authors have identified the big ideas that are at the heart of each book's topic. A set of essential understandings— mathematical points that capture the essence of the topic—fleshes out each big idea. Taken collectively, the big ideas and essential understandings give a view of a mathematics that is focused, connected, and useful to teachers. Links to topics that students encounter earlier and later in school mathematics and to instruction and assessment practices illustrate the relevance and importance of a teacher's essential understanding of mathematics.

On behalf of the Board of Directors, I offer sincere thanks and appreciation to everyone who has helped to make this series possible. I extend special thanks to Rose Mary Zbiek for her leadership as series editor. I join the Essential Understanding project team in welcoming you to these books and in wishing you many years of continued enjoyment of learning and teaching mathematics.

Henry Kepner
President, 2008–2010

# Preface

From prekindergarten through grade 12, the school mathematics curriculum includes important topics that are pivotal in students' development. Students who understand these ideas cross smoothly into new mathematical terrain and continue moving forward with assurance.

However, many of these topics have traditionally been challenging to teach as well as learn, and they often prove to be barriers rather than gateways to students' progress. Students who fail to get a solid grounding in them frequently lose momentum and struggle in subsequent work in mathematics and related disciplines.

The Essential Understanding Series identifies such topics at all levels. Teachers who engage students in these topics play critical roles in students' mathematical achievement. Each volume in the series invites teachers who aim to be not just proficient but outstanding in the classroom—teachers like you—to enrich their understanding of one or more of these topics to ensure students' continued development in mathematics.

## How much do you need to know?

To teach these challenging topics effectively, you must draw on a mathematical understanding that is both broad and deep. The challenge is to know considerably more about the topic than you expect your students to know and learn.

Why does your knowledge need to be so extensive? Why must it go above and beyond what you need to teach and your students need to learn? The answer to this question has many parts.

To plan successful learning experiences, you need to understand different models and representations and, in some cases, emerging technologies as you evaluate curriculum materials and create lessons. As you choose and implement learning tasks, you need to know what to emphasize and why those ideas are mathematically important.

While engaging your students in lessons, you must anticipate their perplexities, help them avoid known pitfalls, and recognize and dispel misconceptions. You need to capitalize on unexpected classroom opportunities to make connections among mathematical ideas. If assessment shows that students have not understood the material adequately, you need to know how to address weaknesses that you have identified in their understanding. Your understanding must be sufficiently versatile to allow you to represent the mathematics in different ways to students who don't understand it the first time.

In addition, you need to know where the topic fits in the full span of the mathematics curriculum. You must understand where your students are coming from in their thinking and where they are heading mathematically in the months and years to come.

Accomplishing these tasks in mathematically sound ways is a tall order. A rich understanding of the mathematics supports the varied work of teaching as you guide your students and keep their learning on track.

### How can the Essential Understanding Series help?

The Essential Understanding books offer you an opportunity to delve into the mathematics that you teach and reinforce your content knowledge. They do not include materials for you to use directly with your students, nor do they discuss classroom management, teaching styles, or assessment techniques. Instead, these books focus squarely on issues of mathematical content—the ideas and understanding that you must bring to your preparation, in-class instruction, one-on-one interactions with students, and assessment.

### How do the authors approach the topics?

For each topic, the authors identify "big ideas" and "essential understandings." The big ideas are mathematical statements of overarching concepts that are central to a mathematical topic and link numerous smaller mathematical ideas into coherent wholes. The books call the smaller, more concrete ideas that are associated with each big idea *essential understandings*. They capture aspects of the corresponding big idea and provide evidence of its richness.

The big ideas have tremendous value in mathematics. You can gain an appreciation of the power and worth of these densely packed statements through persistent work with the interrelated essential understandings. Grasping these multiple smaller concepts and through them gaining access to the big ideas can greatly increase your intellectual assets and classroom possibilities.

In your work with mathematical ideas in your role as a teacher, you have probably observed that the essential understandings are often at the heart of the understanding that you need for presenting one of these challenging topics to students. Knowing these ideas very well is critical because they are the mathematical pieces that connect to form each big idea.

### How are the books organized?

Every book in the Essential Understanding Series has the same structure:

- The introduction gives an overview, explaining the reasons for the selection of the particular topic and highlighting some of the differences between what teachers and students need to know about it.

Big ideas and essential understandings are identified by icons in the books.

marks a big idea, and

marks an essential understanding.

- Chapter 1 is the heart of the book, identifying and examining the big ideas and related essential understandings.

- Chapter 2 reconsiders the ideas discussed in chapter 1 in light of their connections with mathematical ideas within the grade band and with other mathematics that the students have encountered earlier or will encounter later in their study of mathematics.

- Chapter 3 wraps up the discussion by considering the challenges that students often face in grasping the necessary concepts related to the topic under discussion. It analyzes the development of their thinking and offers guidance for presenting ideas to them and assessing their understanding.

The discussion of big ideas and essential understandings in chapter 1 is interspersed with questions labeled "Reflect." It is important to pause in your reading to think about these on your own or discuss them with your colleagues. By engaging with the material in this way, you can make the experience of reading the book participatory, interactive, and dynamic.

Reflect questions can also serve as topics of conversation among local groups of teachers or teachers connected electronically in school districts or even between states. Thus, the Reflect items can extend the possibilities for using the books as tools for formal or informal experiences for in-service and preservice teachers, individually or in groups, in or beyond college or university classes.

marks a "Reflect" question that appears on a different page.

### A new perspective

The Essential Understanding Series thus is intended to support you in gaining a deep and broad understanding of mathematics that can benefit your students in many ways. Considering connections between the mathematics under discussion and other mathematics that students encounter earlier and later in the curriculum gives the books unusual depth as well as insight into vertical articulation in school mathematics.

The series appears against the backdrop of *Principles and Standards for School Mathematics* (NCTM 2000), *Curriculum Focal Points for Prekindergarten through Grade 8 Mathematics: A Quest for Coherence* (NCTM 2006), *Focus in High School Mathematics: Reasoning and Sense Making* (NCTM 2009), and the Navigations Series (NCTM 2001–2009). The new books play an important role, supporting the work of these publications by offering content-based professional development.

The other publications, in turn, can flesh out and enrich the new books. After reading this book, for example, you might select hands-on, Standards-based activities from the Navigations books for your students to use to gain insights into the topics that the Essential Understanding books discuss. If you are teaching students in prekindergarten through grade 8, you might apply your deeper understanding as you present material related to the three focal points that Curriculum Focal Points identifies for instruction at your students' level. Or if you are teaching students in grades 9–12, you might use your understanding to enrich the ways in which you can engage students in mathematical reasoning and sense making as presented in *Focus in High School Mathematics.*

An enriched understanding can give you a fresh perspective and infuse new energy into your teaching. We hope that the understanding that you acquire from reading the book will support your efforts as you help your students grasp the ideas that will ensure their mathematical success.

The authors of the current volume express their sincere appreciation to those who reviewed the manuscript at an earlier stage, including Leslie M. Brickner, Joel Haack, Rhoda Inskeep, Gladis Kersaint, Sharon Lewandowski, and Diane Thiessen. The suggestions and insights that they shared were very helpful in shaping the authors' ideas. The authors also extend appreciation and thanks to Rose Zbiek and Ed Rathmell for their support, ideas, patience, and encouragement during the preparation of this book.

# Introduction

This book focuses on ideas about multiplication and division. These are ideas that you need to understand thoroughly and be able to use flexibly to be highly effective in your teaching of mathematics in grades 3–5. The book discusses many mathematical ideas that are common in elementary school curricula, and it assumes that you have had a variety of mathematics experiences that have motivated you to delve into—and move beyond—the mathematics that you expect your students to learn.

The book is designed to engage you with these ideas, helping you to develop an understanding that will guide you in planning and implementing lessons and assessing your students' learning in ways that reflect the full complexity of multiplication and division. A deep, rich understanding of ideas about these operations will enable you to communicate their influence and scope to your students, showing them how these ideas permeate the mathematics that they have encountered—and will continue to encounter—throughout their school mathematics experiences.

The understanding of multiplication and division that you gain from this focused study thus supports the vision of *Principles and Standards for School Mathematics* (NCTM 2000): "Imagine a classroom, a school, or a school district where all students have access to high-quality, engaging mathematics instruction" (p. 3). This vision depends on classroom teachers who "are continually growing as professionals" (p. 3) and routinely engage their students in meaningful experiences that help them learn mathematics with understanding.

## Why Multiplication and Division?

Like the topics of all the volumes in NCTM's Essential Understanding Series, multiplication and division compose a major area of school mathematics that is crucial for students to learn but challenging for teachers to teach. Students in grades 3–5 need to understand these operations well if they are to succeed in these grades and in their subsequent mathematics experiences. Learners often struggle with ideas about multiplication and division. What is the relationship between multiplication and repeated addition, for example? Many students understand multiplication only as repeated addition. The importance of understanding how arithmetic operations relate to one another and the challenge of computing fluently make it essential for teachers of grades 3–5 to understand multiplication and division extremely well themselves.

Your work as a teacher of mathematics in these grades calls for a solid understanding of the mathematics that you—and your school, your district, and your state curriculum—expect your students to learn about multiplication and division. Your work also requires you to know how this mathematics relates to other mathematical ideas that your students will encounter in the lesson at hand, the current school year, and beyond. Rich mathematical understanding guides teachers' decisions in much of their work, such as choosing tasks for a lesson, posing questions, selecting materials, ordering topics and ideas over time, assessing the quality of students' work, and devising ways to challenge and support their thinking.

# Understanding Multiplication and Division

Teachers teach mathematics because they want others to understand it in ways that will contribute to success and satisfaction in school, work, and life. Helping your students develop a robust and lasting understanding of multiplication and division requires that you understand this mathematics deeply. But what does this mean?

It is easy to think that understanding an area of mathematics, such as multiplication and division, means knowing certain facts, being able to solve particular types of problems, and mastering relevant vocabulary. For example, for the upper elementary grades, you are expected to know such facts as "multiplication of whole numbers is a commutative operation." You are expected to be skillful in solving problems that involve multiplying large numbers. Your mathematical vocabulary is assumed to include such terms as *product*, *divisor*, *remainder*, *factor*, and *multiple*.

Obviously, facts, vocabulary, and techniques for solving certain types of problems are not all that you are expected to know about multiplication and division. For example, in your ongoing work with students, you have undoubtedly discovered that you need not only to know common algorithms for multiplication and division but also to be able to follow strategies that your students create.

It is also easy to focus on a very long list of mathematical ideas that all teachers of mathematics in grades 3–5 are expected to know and teach about multiplication and division. Curriculum developers often devise and publish such lists. However important the individual items might be, these lists cannot capture the essence of a rich understanding of the topic. Understanding multiplication and division deeply requires you not only to know important mathematical ideas but also to recognize how these ideas relate to one another. Your understanding continues to grow with experience and as a result of opportunities to embrace new ideas and find new connections among familiar ones.

Furthermore, your understanding of multiplication and division should transcend the content intended for your students. Some of the differences between what you need to know and what you expect them to learn are easy to point out. For instance, your understanding of the topic should include a grasp of the way in which multiplication of two-digit numbers connects with multiplication of binomial expressions—mathematics that students will encounter later but do not yet understand.

Other differences between the understanding that you need to have and the understanding that you expect your students to acquire are less obvious, but your experiences in the classroom have undoubtedly made you aware of them at some level. For example, how many times have you been grateful to have an understanding of multiplication and division that enables you to recognize the merit in a student's unanticipated mathematical question or claim? How many other times have you wondered whether you could be missing such an opportunity or failing to use it to full advantage because of a gap in your knowledge?

As you have almost certainly discovered, knowing and being able to do familiar mathematics are not enough when you're in the classroom. You also need to be able to identify and justify or refute novel claims. These claims and justifications might draw on ideas or techniques that are beyond the mathematical experiences of your students and current curricular expectations for them. For example, you may need to be able to refute the often-asserted, erroneous claim that three or more whole numbers can be added, subtracted, multiplied, or divided in any order. Or you may need to explain to a student why "carrying" is appropriate in some multiplication work.

# Big Ideas and Essential Understandings

Thinking about the many particular ideas that are part of a rich understanding of multiplication and division can be an overwhelming task. Articulating all of those mathematical ideas and their connections would require many books. To choose which ideas to include in this book, the authors considered a critical question: What is essential for teachers of mathematics in grades 3–5 to know about multiplication and division to be effective in the classroom? To answer this question, the authors drew on a variety of resources, including personal experiences, the expertise of colleagues in mathematics and mathematics education, and the reactions of reviewers and professional development providers, as well as ideas from curricular materials and research on mathematics learning and teaching.

As a result, the mathematical content of this book focuses on essential ideas for teachers about multiplication and division. In particular, chapter 1 is organized around two big ideas related to this important area of mathematics. Each big idea is supported by smaller, more specific mathematical ideas, which the book calls *essential understandings*.

# Benefits for Teaching, Learning, and Assessing

Understanding multiplication and division can help you implement the Teaching Principle enunciated in *Principles and Standards for School Mathematics*. This Principle sets a high standard for instruction: "Effective mathematics teaching requires understanding what students know and need to learn and then challenging and supporting them to learn it well" (NCTM 2000, p. 16). As in teaching about other critical topics in mathematics, teaching about multiplication and division requires knowledge that goes "beyond what most teachers experience in standard preservice mathematics courses" (p. 17).

Chapter 1 comes into play at this point, offering an overview of multiplication and division that is intended to be more focused and comprehensive than many discussions of the topic that you are likely to have encountered. This chapter enumerates, expands on, and gives examples of the big ideas and essential understandings related to the operations, with the goal of supplementing or reinforcing your understanding. Thus, chapter 1 aims to prepare you to implement the Teaching Principle fully as you provide the support and challenge that your students need for robust learning about multiplication and division.

Consolidating your understanding in this way also prepares you to implement the Learning Principle outlined in *Principles and Standards*: "Students must learn mathematics with understanding, actively building new knowledge from experience and prior knowledge" (NCTM 2000, p. 20). To support your efforts to help your students learn about multiplication and division in this way, chapter 2 builds on the understanding of these operations that chapter 1 communicates by pointing out specific ways in which the big ideas and essential understandings connect with mathematics that students typically encounter earlier or later in school. This chapter supports the Learning Principle by emphasizing longitudinal connections in students' learning about multiplication and division. For example, as their mathematical experiences expand, students gradually develop an understanding of the connections between multiplication and division of whole numbers and the corresponding operations with fractions and decimals and become fluent in using properties of operations and numbers in estimation and mental mathematics.

The understanding that chapters 1 and 2 convey can strengthen another critical area of teaching. Chapter 3 addresses this area, building on the first two chapters to show how an understanding of multiplication and division can help you select and develop appropriate tasks, techniques, and tools for assessing your students' understanding of the operations. An ownership of the big ideas and essential understandings related to multiplication and division, reinforced by an understanding of students' past and future experiences with the ideas, can help you ensure that assessment in your classroom supports the learning of significant mathematics.

Such assessment satisfies the first requirement of the Assessment Principle set out in *Principles and Standards*: "Assessment should support the learning of important mathematics and furnish useful information to both teachers and students" (NCTM 2000, p. 22). An understanding of multiplication and division can also help you satisfy the second requirement of the Assessment Principle, by enabling you to develop assessment tasks that give you specific information about what your students are thinking and what they understand. For example, when asked to multiply two two-digit numbers such as 24 and 35, a student might write something like

$$\begin{array}{r} {\scriptstyle 2} \\ 35 \\ \times\ 24 \\ \hline 140 \end{array}$$

while saying "4 times 5 is 20, so I put down the zero and carry the 2; 4 times 3 is 12 and 2 more is 14, so I write down 140; and then...." The student's response to the question, "Why do you add 12 and 2 when this is a multiplication problem?" can reveal much about how the student understands the mechanics of a common algorithm that he or she will soon extend to products of numbers with more than two digits.

## Ready to Begin

This introduction has painted the background, preparing you for the big ideas and associated essential understandings related to multiplication and division that you will encounter and explore in chapter 1. Reading the chapters in the order in which they appear can be a very useful way to approach the book. Read chapter 1 in more than one sitting, allowing time for reflection. Absorb the ideas—both big ideas and essential understandings—related to multiplication and division. Appreciate the connections among these ideas. Carry your newfound or reinforced understanding to chapter 2, which guides you in seeing how the ideas related to these operations are connected to the mathematics that your students have encountered earlier or will encounter later in school. Then read about teaching, learning, and assessment issues in chapter 3.

Alternatively, you may want to take a look at chapter 3 before engaging with the mathematical ideas in chapters 1 and 2. Having the challenges of teaching, learning, and assessment issues clearly in mind, along with possible approaches to them, can give you a different perspective on the material in the earlier chapters.

No matter how you read the book, let it serve as a tool to expand your understanding, application, and enjoyment of multiplication and division.

# Multiplication and Division: The Big Ideas and Essential Understandings

As *Principles and Standards for School Mathematics* (NCTM 2000) asserts, "Developing number sense, understanding number and operations, and gaining fluency in arithmetic operations form the core of mathematics education for the elementary grades" (p. 32). Furthermore, establishing a deep understanding of multiplication and division is a major expectation in the upper elementary years. *Curriculum Focal Points for Prekindergarten through Grade 8 Mathematics: A Quest for Coherence* (NCTM 2006) emphasizes the importance of understanding multiplication and division in grades 3–5, targeting this understanding as a focal point for instruction in each of these grades. In its final report, *Foundations of Success*, the National Mathematics Advisory Panel (2008) recommends that instruction help elementary students develop a firm understanding of the meanings of the operations of multiplication and division as well as computational fluency with them. As described in all three of these influential documents, understanding involves much more than just computational fluency.

The purpose of this book is to identify and describe many of the ideas associated with multiplication and division of whole numbers and rational numbers. Thus, the book will examine multiple—

- situations for which the operations of multiplication and division are appropriate choices to solve problems;

- ways to represent multiplication and division;

- ways to reason with multiplication and division; and

- connections and relationships among these and other mathematical topics.

It will also examine numerical relationships that result from multiple representations and the reasoning required for the meaningful use and understanding of computational algorithms, written and mental, standard and nonstandard.

"Unpacking" the ideas related to multiplication and division is a critical step in developing a deeper understanding. To those without specialized training, many of these ideas might appear to be easy to teach. But those who teach students in grades 3–5 are aware of their subtleties and complexities.

Not only is the operation of multiplication inextricably connected with the operation of division, but multiplication also plays an integral role in many other significant mathematical topics. These include prime and composite numbers, factorization and prime factorization, factor and greatest common factor, multiple and least common multiple, area and volume, proportional reasoning, mean, algebraic expression, linear function, and place value.

This chapter organizes this mathematical content by relating it to two big ideas about multiplication and division. The first big idea involves different problem structures and representations related to multiplication and division and the connections and numerical relationships among them. The second big idea focuses on the reasoning required for performing and understanding computational algorithms for multiplication and division. Each of these big ideas involves several smaller, more specific "essential understandings." The two big ideas and all the associated understandings are identified as a group below to give you a quick overview and for your convenience in referring back to them later. Read through them now, but do not think that you must absorb them fully at this point. The chapter will discuss each one in turn in detail.

**Big Idea 1.** Multiplication is one of two fundamental operations, along with addition, which can be defined so that it is an appropriate choice for representing and solving problems in many different situations.

> **Essential Understanding 1a.** In the multiplicative expression $A \times B$, $A$ can be defined as a *scaling factor*.

> **Essential Understanding 1b.** Each multiplicative expression developed in the context of a problem situation has an accompanying explanation, and different representations and ways of reasoning about a situation can lead to different expressions or equations.

> **Essential Understanding 1c.** A situation that can be represented by multiplication has an element that represents the scalar and an element that represents the quantity to which the scalar applies.

Essential Understanding 1*d*. A scalar definition of multiplication is useful in representing and solving problems beyond whole number multiplication and division.

Essential Understanding 1*e*. Division is defined by its inverse relationship with multiplication.

Essential Understanding 1*f*. Using proper terminology and understanding the division algorithm provide the basis for understanding how numbers such as the *quotient* and the *remainder* are used in a division situation.

## Big Idea 2. The properties of multiplication and addition provide the mathematical foundation for understanding computational procedures for multiplication and division, including mental computation and estimation strategies, invented algorithms, and standard algorithms.

Essential Understanding 2*a*. The commutative and associative properties of multiplication and the distributive property of multiplication over addition ensure flexibility in computations with whole numbers and provide justifications for sequences of computations with them.

Essential Understanding 2*b*. The right distributive property of division over addition allows computing flexibly and justifying computations with whole numbers, but there is no left distributive property of division over addition and no commutative or associative property of division of whole numbers.

Essential Understanding 2*c*. *Order of operations* is a set of conventions that eliminates ambiguity in, and multiple values for, numerical expressions involving multiple operations.

Essential Understanding 2*d*. Properties of operations on whole numbers justify written and mental computational algorithms, standard and invented.

# Multiplication Is a Powerful Tool: Big Idea 1

---

Big Idea 1. *Multiplication is one of two fundamental operations, along with addition, which can be defined so that it is an appropriate choice for representing and solving problems in many different situations.*

---

Multiplication is a fundamental operation that is used to solve everyday problems. Considering examples makes it easy to understand why multiplication is an appropriate choice in many different problem situations. In fact, research suggests that beginning with problem situations helps students develop competence in computation and problem solving (Fuson 2003).

## Multiplication as a scalar operation

---

Essential Understanding 1a. *In the multiplicative expression $A \times B$, A can be defined as a* scaling factor.

---

Multiplication has been described by using repeated addition, area, a Cartesian product, and a rectangular array, among other possibilities. Students' understanding of multiplication is enhanced when they have opportunities to think about and model it in various ways. Phrases such as "twice as many cups," "one-half of that pizza," or "five times as much money" motivate students to develop a concept of multiplication that builds on their informal understanding of these situations and acts as a touchstone to help them interpret other descriptions of multiplication.

Consider the problem in example 1, set in the familiar context of measuring ingredients according to a recipe:

> Example 1: A recipe calls for 2 cups of flour for a cake. How many cups of flour will we need if we are going to make 5 cakes?

We know that the quantity of flour that we will need for five cakes is five times as great as the quantity of flour that we would need for one cake, which requires only two cups. Or more simply, we will need five times as much flour, and we recognize this as a situation in which we would usually use multiplication and write $5 \times 2$. Figure 1.1 is a pictorial representation of the quantity of flour needed.

Although it is easy to show students how we picture a situation, we learn a great deal about how they understand the quantities and operations involved in the situation when they create their own representations of problems (Quintero 1986). This information can

Fig. 1.1. Five times as much flour as the two cups for one cake

serve as a guide in decisions about a teaching strategy and appropriate materials. Reflect 1.1 probes this idea.

### Reflect 1.1

What pictorial figures might your students draw to represent the situation in example 1?

If possible, give your students the problem and ask them to illustrate the situation, or ask colleagues to give the problem to their students to illustrate it with a picture.

Consider differences among the illustrations. What do the differences mean, and how might they shape instruction?

In example 1, we know that we will need five times as much flour for five cakes as for one cake. The first factor, 5, is called a *scaling factor*, or *scalar factor*, *scale factor*, or, sometimes, *scalar operator*. The first factor will "scale" the second quantity; that is, it will resize the second factor proportionally according to the scale given by the first factor, thus making a multiplicative change, as contrasted with an additive change. The impact of a different scalar factor—say, $1/2$—can be envisioned in much the same way, except that $1/2$ changes the second quantity to half as much. In this case, multiplication is used as a scalar operation.

The examples that follow present several problem situations in which we might choose the operation of multiplication or division as our solution strategy. The discussion of these examples highlights points about how we—

- use multiplication as a scalar operation;
- represent a situation in more than one way;
- discover connections among the various representations;
- examine connections between multiplication and division; and
- justify a choice of an operation on the basis of the meaning of that operation.

The goal of the discussion is to make sense of the concepts connected with multiplication and division. Thoughtful study of these examples, with an emphasis on the five points above, can deepen understanding of multiplication and division. The question in Reflect 1.2 foreshadows ways of reasoning about a situation.

## Reflect 1.2

What operations might someone choose to use in converting a person's weight measured in pounds as 156 pounds into a weight in kilograms. Why?

*Note:* One pound is approximately equivalent to 0.454 kilograms.

## Reasoning about situations and representations

*Essential Understanding 1b. Each multiplicative expression developed in the context of a problem situation has an accompanying explanation, and different representations and ways of reasoning about a situation can lead to different expressions or equations.*

The expression that we use to model a situation reflects our reasoning, and our explanation of our thinking guides our way of representing the situation and any expression or equation that we write. What multiplicative expressions might model the situation in example 2 below? What explanations and representations might accompany them?

> **Example 2:** If each of Robert's shirts has 4 buttons, then how many buttons are on 5 of his shirts?

One representation of this situation could be a drawing of 5 shirts, each with 4 buttons. How would your students count the number of buttons on the 5 shirts? A picture of one shirt with 4 buttons appears in figure 1.2.

**Fig. 1.2. One shirt with 4 buttons**

One solution strategy would be to consider that the situation presents a repeated addition problem:

$$4 + 4 + 4 + 4 + 4 = \square$$

To explain the addition expression, we would say that the first 4 represents the number of buttons on the first shirt, the second 4 represents the number of buttons on the second shirt, and so on, with the number of terms, or addends (5), representing the number of shirts. A common description of such a situation is "5 groups of 4 objects," where the objects are the buttons on *one* shirt.

We might also write this sum as the multiplicative expression $5 \times 4$. Multiplication is sometimes described as repeated addition. Unfortunately, this description can create some rather awkward interpretations in other situations—for example, with fractions, as chapter 2 will discuss.

Under an interpretation of multiplication as a scalar operation, the total number of buttons is five times as many as the number of buttons on one shirt. We could say that the 5 is prescribing the multiplicative action that we should make with the number of buttons on the one shirt. We want 5 times as many buttons. Note that we have not calculated the numerical value of $5 \times 4$, since our focus is on the meanings of the factors 5 and 4.

We could also represent the situation by using the multiplicative expression $4 \times 5$, without using the commutative property, by noting that we have 4 groups of 5 objects. We have 5 first buttons on the 4 shirts, 5 second buttons on the 4 shirts, and so on, for the other two buttons. Interpreting the situation in this way, as "4 groups of 5 objects," might not be as natural a choice as interpreting it in the first way, as "5 groups of 4."

An explanation is always necessary to clarify the reasoning behind an expression or equation. Note that in the first interpretation of the problem, as 5 groups of 4 objects, with the accompanying expression $5 \times 4$, we have been very careful to use 5 to describe the number of shirts, not the shirts themselves. Similarly, we have used 4 to refer to the number of buttons on a shirt, not the buttons themselves. In the second interpretation of the problem, as 4 groups of 5 objects, with the accompanying expression $4 \times 5$, 4 represents the number of buttons, not the buttons themselves, and 5 describes, for example, the number of first buttons, not the first buttons themselves.

Making this kind of distinction is an important part of establishing understanding in mathematics, especially algebra. Other numbers besides 4 might be associated with the buttons. For example, 25 might be their cost in cents, and 2 might be their weight in ounces. Dougherty and colleagues (2010) offer an extended discussion of the important difference between objects and quantities.

A picture showing 5 shirts, each with 4 buttons, is not, of course, the only possible representation of the situation. Figure 1.3 shows a table with five rows, each of which represents a shirt, and four columns, each of which represents a button. In the table, each cell represents a certain button on a certain shirt. For example, the orange cell in the figure represents the third button on the second shirt. The rows and columns of this table represent the situation in example 2 as an *array*.

For an extended discussion of how different quantities can be associated with an object, see *Developing Essential Understanding of Number and Numeration for Teaching Mathematics in Prekindergarten–Grade 2* (Dougherty et al. 2010).

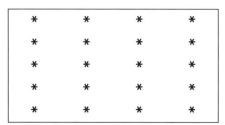

| shirt\button | button 1 | button 2 | button 3 | button 4 |
|---|---|---|---|---|
| shirt 1 | | | | |
| shirt 2 | | | shirt 2 button 3 | |
| shirt 3 | | | | |
| shirt 4 | | | | |
| shirt 5 | | | | |

Fig. 1.3. A table representing five shirts, each with four buttons

The use of an array as a representation is more familiar in the context of a situation like that in example 3:

**Example 3:** A classroom has a rectangular arrangement of desks with 5 rows of 4 desks. How many desks are in the classroom?

A natural pictorial representation would be a rectangular array with 5 horizontal rows, with each row repeating the same symbol four times, to represent the desks in the row. Figure 1.4 shows one example. Reflect 1.3 provides an opportunity to make sense of the representation and situation.

```
*     *     *     *

*     *     *     *

*     *     *     *

*     *     *     *

*     *     *     *
```

Fig. 1.4. Representing 5 rows of 4 desks as an array

## Reflect 1.3

How does 4 × 5 make sense as an answer to the question in example 3?

How does 5 × 4 make sense as an answer to the question?

# Representing situations with multiplication and division

Essential Understanding 1c. *A situation that can be represented by multiplication has an element that represents the scalar and an element that represents the quantity to which the scalar applies.*

Problem solvers often choose the operation of division for use in solving a problem such as example 4:

**Example 4:** Grandmother has 21 marbles that she wants to give to her grandchildren. Each grandchild receives 7 marbles. How many grandchildren does Grandmother have?

With a division interpretation, dividing 21 by 7 gives the answer, or *quotient*. This type of situation is frequently called a *measurement* model, or *repeated subtraction* model, for division. The word *measurement* is used because the problem is sometimes restated as, "How many groups of 7 marbles are in a group of 21 marbles?" Answering this question involves "measuring" how many. Figure 1.5 shows a pictorial and a symbolic representation that are part of the solution strategies of two students. How are these representations similar? Can you find another strategy besides division?

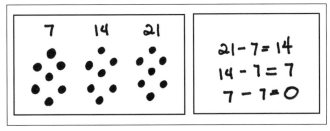

Fig. 1.5. Two students' written work for "How many groups of 7 marbles are in a group of 21 marbles?"

Why is division an appropriate choice of operation for use in solving the problem in example 4? Students frequently choose an operation to solve a problem without trying to make sense of the choice. They may look for key words or phrases—for example, "It says 'how many in each group,' so I must divide." They may just look at the numbers and choose an operation on the basis of their size—for instance, "There is one big number and one small number, so I must divide." They may expect an answer that is less than one of the numbers—for example, "I know the answer is smaller, so I have to divide." Or they may note that the numbers are compatible—that is, one number is a multiple of the other—for instance, "Seven divides 21, so I must divide." Reasons such as these are never valid. Knowing why an operation is an appropriate choice for a solution strategy is an important part of establishing a robust understanding of mathematics.

The situation in example 4 can also be interpreted as multiplication. Each grandchild receives 7 marbles. What number—call it $A$—describes how many times greater the total number of marbles is compared with the number of marbles that each grandchild gets? This means that there are $A$ groups of 7 marbles, with a total of 21 marbles, or, equivalently, that $A$ is a number such that $A \times 7 = 21$. Understanding multiplication as a scalar operation with $A$ as the scalar allows us to transform a division problem into one that we can solve by using the operation of multiplication.

The use of multiplication where division can also be used involves applying the *missing factor* interpretation of division. The choice of division as the operation for these multiplication situations will be described later.

Contrast the new situation in example 5 with the previous one in example 4:

**Example 5:** Grandfather has 24 old coins. He is going to give each of his 6 grandchildren the same number of coins. How many will each grandchild receive?

Problems like this are often modeled by division, reflecting the *sharing*, or *partitive*, interpretation of the operation. In example 5, division can be used to obtain the answer when 24 is divided by 6. Reflect 1.4 explores different ways in which children might think about this problem.

## Reflect 1.4

What are two different ways in which a child might think about the problem in example 5 to solve it?

Much as we did in the case of example 4, we can interpret the situation in example 5 as multiplicative. The total number of grandchildren, 6, indicates that the total number of coins, 24, is 6 times as many as the number of coins that each grandchild will receive. If we call the number of coins that each grandchild will get $B$, then we have 6 groups of $B$ coins with a total of 24 coins. We can write this as $6 \times B = 24$, with 6 as the scalar. So again, interpreting multiplication as a scalar operation allows us to transform a division situation into a multiplicative situation. Again, we have a missing factor interpretation of division.

The last two examples illustrate how the operations of multiplication and division can each be used to represent the same situation. It is always important that the representation—whether it be pictorial or an arithmetic expression or an equation—exactly follow the reasoning about the situation. Understanding connections among different representations for a single situation helps develop deeper understanding.

Example 6 presents a different type of problem:

**Example 6:** How many different outfits can Sue make if she has 3 shirts (white, red, and blue) and 2 skirts (tan and navy)?

Problems of this sort are frequently described as *counting* or *combinatorics* problems. Hendrickson (1986) refers to such a problem as a *selection problem* and contrasts it with other types of multiplication and division problems that students encounter. It is important to be able to recognize and represent a problem of this sort as a multiplication situation and to compare it with problems of the other types illustrated in this book. To problem solvers, the value of doing so is similar to the value of knowing the different types of addition and subtraction problems discussed by Caldwell and colleagues (2011).

The problem about Sue's outfits involves counting the number of ways in which a different outfit can be selected. Multiplication is frequently used to find the answer—in this case, by finding the product $3 \times 2$, if the multiplication principle for combinatorics is mimicked exactly. But why does multiplication make sense? A tree diagram like that in figure 1.6 can model this situation. Starting on the left, we see three choices for a shirt: white, red, or blue. The path from each of those choices leads to another choice for a skirt: tan or navy. Every choice of a color of a shirt and a color of a skirt is represented by one of the six paths from the starting point, resulting in six possible outfits.

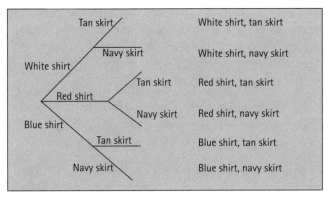

**Fig. 1.6. A tree diagram representing all six possible outfits**

How can this representation be described by multiplication? For each shirt, Sue has two ways to complete the outfit. Because she has three choices for the color of the shirt, she has three times as many different outfits as the number of skirts. Reasoning in this way supports viewing 3 as a scalar operator acting on the number of skirts, resulting in the multiplicative expression $3 \times 2$. Consider how you might draw a tree to illustrate representing the problem by the multiplicative expression $2 \times 3$.

A tree diagram is not the only possible choice to represent the shirt-skirt counting situation in example 6. A table creates a different representation, as shown in figure 1.7. Each row indicates the color of the shirt, and each column indicates the color of the skirt. The shaded cell represents an outfit consisting of a white shirt with a navy skirt. Furthermore, the six cells represent all possible outfits, an observation that is necessary to complete the solution.

For a discussion of different types of addition and subtraction problems, see *Developing Essential Understanding of Addition and Subtraction for Teaching Mathematics in Prekindergarten–Grade 2* (Caldwell et al. 2011).

|        | tan | navy |
|--------|-----|------|
| white  |     |      |
| red    |     |      |
| blue   |     |      |

Fig. 1.7. A table representing all six possible outfits

Using reasoning similar to that used with example 3, we can see that this situation can be interpreted by using scalar multiplication, resulting in an expression of $3 \times 2$. Also consider how to use scalar multiplication to interpret the table representation in a way that results in an answer of $2 \times 3$.

By using ordered pairs, we can make yet another representation of the situation in example 6. The first component in each ordered pair can represent the color of the shirt, and the second component can represent the color of the skirt. Our representation will list all possible combinations. Using a systematic approach, we might produce the following as one listing of ordered pairs:

| | |
|---|---|
| (white, tan) | (white, navy) |
| (red, tan) | (red, navy) |
| (blue, tan) | (blue, navy) |

A comparison of the ordered-pair representation with the table representation shows a connection between these two representations: two ordered pairs can be constructed for each row in the table by using the row name as the first component and the column name as the second component. Like the table representation, the ordered-pair representation gives rise to a natural interpretation of example 6 as a scalar multiplicative situation, with the expression $3 \times 2$ leading to the answer. In selecting an outfit, Sue has 2 choices of skirts to go with each shirt. Because she has 3 choices of shirts, she has 3 times as many choices of outfits as her choices of skirts, or $3 \times 2$ choices in all.

The use of ordered pairs in this manner is frequently called the *Cartesian product* model of multiplication. Formally, the Cartesian product of two sets $A$ and $B$ is created by making the set of all ordered pairs where the first component is an element of $A$ and the second component is an element of $B$. The product of two whole numbers $a$ and $b$ can be defined as the number of elements in the Cartesian product of $A$ and $B$, where $a$ is the number of elements in the set $A$ and $b$ is the number of elements in the set $B$. Reflect 1.5 explores the Cartesian product model and relationships among it, the tree diagram, and the table in the case of example 6.

See Reflect 1.5 on p. 19.

**Reflect 1.5**

Find 3 × 4 by using the Cartesian product representation.

We have discussed three representations in connection with example 6—a tree diagram, a table, and a Cartesian product model. How are these three representations similar? How are they different?

Example 7 presents a problem that involves finding the area of a rectangular region:

**Example 7:** A rectangular carpet measures 5 feet by 4 feet. How many square feet of floor will the carpet cover?

A typical strategy for solving this problem would be to multiply the length and the width to get an answer of 20 square feet. Why is this situation multiplicative?

We might represent the carpet situation by drawing a rectangle in which one side has a length of 5 units and another side has a length of 4 units, as shown in figure 1.8. Then we might partition the rectangle into unit squares, and so the answer would be the number of unit squares, or, in example 7, 20 square feet.

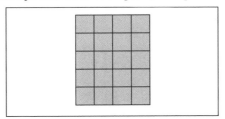

**Fig. 1.8. A rectangle representing carpet that is 5 feet by 4 feet**

Consider only the first row of the representation. It is a rectangle with one row of unit squares. This first row with 4 unit squares has area 4. The entire representation includes 5 rows, which means there are 5 times as many rows, or, equivalently, 5 times as many unit squares as in the first row. This is a multiplicative situation in which 5 is a scalar operator indicating that we want 5 times as much area as the area of the first row. The number of unit squares, or area, is expressed by 5 × 4.

## Moving beyond operations on whole numbers

*Essential Understanding 1d. A scalar definition of multiplication is useful in representing and solving problems beyond whole number multiplication and division.*

A scalar interpretation of multiplication can give students broad and flexible understanding of multiplication and division and help them reason about and explain their use of these operations in problem solving. Example 8 offers a starting point for examining this idea:

**Example 8:** Write an algebraic expression to show that Tom has twice as many pounds of candy as Susan.

Representing the situation with a diagram, followed by carefully and precisely defining the variables or components of that diagram, is a good opening strategy. When students attempt to use algebraic strategies to solve word problems, a failure to describe variables carefully at the start often dooms their efforts. For example, $T$ can represent the number of pounds of candy that Tom has, and $S$ can represent the number of pounds of candy that Susan has, but students sometimes fail to note that both $T$ and $S$ represent numbers of pounds—not the candy itself. A solution needs to describe the relationship between the amount of candy that Tom has and the amount of candy that Susan has.

Students often attempt to translate the words of a problem directly into a symbolic expression. In the case of example 8, this might lead them to write $2T = S$. Interpreting the multiplication expression $2T$ by taking 2 as a scalar factor indicates that if the amount of candy that Tom has is doubled, then this doubled amount would be the same amount of candy that Susan has. But this is obviously not the situation described in the example.

The diagram in figure 1.9 shows the relationship between the amounts of candy that Susan and Tom have. The first figure represents the amount of candy Susan has ($S$), and the second figure represents the amount of candy Tom has ($T$). Now it is clear that if the amount that Susan has is doubled, as represented by $2S$, it would then be the same as the amount that Tom has. Hence, the relationship between the amounts of candy that Tom and Susan have is $T = 2S$.

Fig. 1.9. A diagram showing that Tom has twice as many pounds of candy as Susan

Multiplication as a scalar operation provides a way of thinking about the situation and a justification for the choice of multiplication. Again, the importance of providing a justification for the choice of an operation is an important part of understanding and doing mathematics. Furthermore, the advantage of establishing a robust understanding of the problem, such as that provided by a pictorial representation, rather than beginning with the manipulation of algebraic symbols, is apparent.

Example 9 offers another glimpse of the value of the scalar interpretation in reasoning about and explaining the choice of operation in a solution strategy:

**Example 9:** Jose drives at an average speed of 50 miles per hour for 6 hours. How many miles does Jose drive?

This situation is generally recognized as a multiplicative one calling for the use of the distance = rate $\times$ time formula, or $d = rt$. Following the order of the letters labeling the variables in the formula, most students would write $50 \times 6$ as the multiplication expression that will give the solution. But the format of this answer suggests that we are looking at 50 groups of 6 objects. What would each object be? An hour? Also, why do we use the operation of multiplication? Examining this problem from a different perspective offers an alternate way to make sense of the situation.

Jose drives at an average a rate of 50 miles per hour, which we can think of as meaning that Jose travels 50 miles in one hour and in each hour thereafter. What happens when Jose drives 6 times as much as he does in one hour? As suggested by the previous example, a pictorial representation can establish an understanding of the problem. In figure 1.10, each line segment represents 50 miles, which is the distance Jose travels in one hour, and the six segments represent Jose's travel in six hours. Multiplication as a scalar operation applies when the 6 is interpreted as a scalar describing how the quantity 50 is changed, yielding a multiplicative expression of $6 \times 50$. The answer and the initial quantity—the quantity to which the scalar applies—have the same units—namely, miles.

Fig. 1.10. Six segments, with each representing 50 miles traveled in one hour

We can use multiplication to develop and generalize a relationship among distance, rate, and time. We begin by defining the variables carefully and precisely:

$D$ represents the total number of miles traveled.
$T$ represents the number of hours used in traveling $D$ miles.
$R$ represents the number of miles traveled in each hour.

One way to make sense of this general situation is to reason as before, using a scalar view of multiplication. If we consider $T$ as a scalar acting on $R$, the relationship makes sense. $R$ is the number of miles traveled each hour, so $T \times R$ yields the number of miles traveled in $T$ hours. This interpretation of the multiplicative situation leads to the algebraic formula $D = T \times R$. In any rate problem involving multiplication, it is then critical to consider what is being used as the scalar.

Every choice of an operation requires a justification: why is this operation appropriate? An understanding of the meaning of the operation is demonstrated in this way. Consider the following two examples:

**Example 10:** If an object weighs 156 pounds, and one pound has approximately the same weight as 0.454 kilograms, then what is an approximate weight of the object in kilograms?

This problem previously appeared in Reflect 1.2. Can you recall your ideas and reasoning as you considered the problem there? Compare your earlier thinking with the ideas that a scalar interpretation has given you now.

Example 11 presents a problem for which the operation of choice would probably be either multiplication or addition:

**Example 11:** Suppose that you are writing the first few terms in the sequence 1, 3, 6, 10, ... of triangular numbers. What is the next term that you will write?

The choice of an operation should be based on an understanding of the situation—in this case, the structure of the triangular numbers—and the meaning of the chosen operation. Looking at some noncontextual pattern or using some memorized technique, such as dimensional analysis, should not be a basis for the choice. Knowing which operation to use and why it is appropriate is an indicator of understanding.

A common aim in the previous examples has been to see how describing multiplication as a scalar operation makes sense for a variety of reasons. In a multiplicative expression, the first factor describes the action that is to happen to the quantity described by the second factor. The first number in the product will change, or *scale*, the quantity described by the second number in the expression. The change is multiplicative. This description is powerful and helps students to develop a robust understanding of multiplication. As a result of applying this interpretation, problem solvers can—

- describe many varied situations in a multiplicative manner;
- use multiple representations to describe multiplicative situations;
- provide justifications for using multiplication as an operation;
- make connections among different representations;
- trace the "answer," or product, of a multiplicative expression to the interpretation of a representation; and
- give meanings to the factors in a multiplicative expression.

# Relating division and multiplication

Essential Understanding 1e. *Division is defined by its inverse relationship with multiplication.*

So far, our discussion has centered on multiplication and on a way to describe this operation that shows that it is an appropriate choice in many different situations. The discussions of example 4, in which Grandmother has 21 marbles and gives 7 to each of her grandchildren, and example 5, in which Grandfather has 24 old coins to share equally among his 6 grandchildren, make brief references to the use of division in finding the number of equal-sized groups and the number of items in equal-sized groups. These examples present situations in which the operation of division can be used. But that observation in itself does not give a meaning for the operation of division that can be used to justify the choice of division.

Consider the relationship between addition and subtraction. When $b$ is subtracted from $a$, the result is represented by $a - b$ and is defined to be the number $y$ where $a = b + y$. Addition and subtraction have an inverse relationship. In other words, addition is a fundamental operation and subtraction is defined in terms of addition. Likewise, multiplication is a fundamental operation, with division defined in terms of multiplication.

We give meaning to division in a manner that is analogous to the way in which we give meaning to subtraction. With $b \neq 0$, the result of dividing $a$ by $b$ is represented by $a \div b$ and is defined to be the number $y$ where $a = b \times y$, or, equivalently, $y = a \times (1/b)$ where $1/b$ is the notation for the multiplicative inverse of $b$. Note the similar situation with addition and subtraction, in which we know that $a - b = a + (-b)$ where $-b$ is the notation for the additive inverse of $a$. The first format gives the connection of division with multiplication, so multiplication and division have an inverse relationship. The equation $a = b \times y$ describes the same relationships among the numbers as the equation $a \div b = y$. The expression $a \times (1/b)$ is often written as $a(1/b)$, or $a/b$, a connection with fractions that is consistent with the way in which fractions are used.

The inverse relationship between multiplication and division provides the mathematical basis for the fact families, such as the following:

$7 \times 3 = 21$     $3 \times 7 = 21$     $21 \div 3 = 7$     $21 \div 7 = 3.$

Which of the four facts above we use depends on the situation and our reasoning for the choice of the operation. Chapter 3 discusses this thinking in more detail, and Caldwell and colleagues (2011) discuss similar thinking in relation to addition and subtraction.

In example 4, where Grandmother has 21 marbles and gives 7 to each grandchild, problem solvers must determine how many

For a discussion of the usefulness of addition fact families, see *Developing Essential Understanding of Addition and Subtraction for Teaching Mathematics in Prekindergarten– Grade 2* (Caldwell et al. 2011).

grandchildren Grandmother has. As we noted earlier, problem solvers frequently choose to use division in solving this kind of problem, but often without being able to justify the choice. Indeed, when we discussed this problem, we had not yet discussed the mathematical reasons that justify the choice of division, so instead we employed a strategy using multiplication to determine the number of grandchildren, $A$, by developing the algebraic equation $A \times 7 = 21$. Now, however, by using our recognition of the inverse relationship between multiplication and division, we can write the equation $A = 21 \times (1/7)$. This equation indicates that we could have obtained the answer by using division, dividing 21 by 7. Thus, the choice of division is fully appropriate, since we have now given a justification for it. Reflect 1.6 probes the use of similar reasoning to justify the choice of division in the case of example 5.

## Reflect 1.6

Explain why division can be an appropriate choice of operation to use in solving the problem in example 5:

Grandfather has 24 old coins. He is going to give each of his 6 grandchildren the same number of coins. How many will each grandchild receive?

## Using numbers in division situations

Essential Understanding 1f. *Using proper terminology and understanding the division algorithm provide the basis for understanding how numbers such as the* quotient *and the* remainder *are used in a division situation.*

Often the terminology that is used with division is vague, inappropriate, or misleading. It can even interfere with the development of a rich understanding of division. Proper use of terminology and symbols can increase understanding of the meanings of arithmetic expressions and help to avoid mistakes. Consider the following examples:

- "Suppose that a group of apples is divided in half." The meaning of this statement is ambiguous, as illustrated in parts (a) and (b) of figure 1.11. Does it mean that the group of apples is partitioned into two equal-sized groups of whole apples, as shown in (a), or does it mean that each apple is partitioned into two same-sized pieces, as shown in (b)? These are two very different situations, as the figure suggests.

- "How many times does 6 go into 42?" The use of the phrase "goes into" is frequently used in conjunction with division,

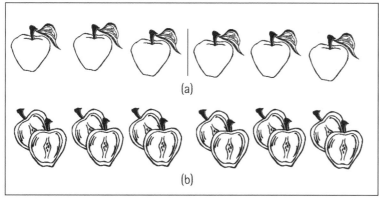

Fig. 1.11. Six apples (a) partitioned into two equal-sized groups of whole apples and (b) with each apple partitioned into two same-size pieces

particularly long division. This phrase really has no mathematical meaning, so it should be avoided.

- "Suppose that 6 is divided into 12." This phrase lends itself to various interpretations. Does it mean to divide 6 by 12 or 12 by 6? Does it mean to do the division, or does it refer to the answer? Phrases that can be interpreted in different ways should be avoided; otherwise the listener—usually a student—can interpret them in ways that are not intended.

The use of correct notation to represent some aspect of a problem is often a very important first step. For example, students frequently encounter statements such as, "The natural number $a$ is divisible by 3." They may then proceed to represent this statement by the expression $a/3$, which is an incorrect representation and often makes finding a solution strategy very difficult, if one can be found at all. It would appear that students are confusing the *result* of dividing $a$ by 3 with the *operation* of dividing $a$ by 3. These are two separate ideas. In general, when $b$ is a nonzero divisor, it is appropriate to say "divide $a$ by $b$," with the result written as $a/b$. Note that $a/b$ is an answer to a division problem and does not mean to divide $a$ by $b$ or that $a$ is divisible by $b$. The operation of division has already been carried out when we see an expression such as $a/b$. Explaining answers to the questions in Reflect 1.7 can help to connect notation and terminology for division and multiplication.

## Reflect 1.7

We can say "divide $a$ by $b$."
Is it appropriate to say "divide $a$ and $b$"?
Does it make sense to say "multiply $a \times b$"?
Does it make sense to say "divide $a/b$"?

It is critical that mathematical ideas be communicated carefully and stated precisely to promote an understanding of concepts, relationships, and representations. Mathematical terms are well defined and should be used instead of words and phrases that do not have mathematical meanings.

The use of the term *divisible* involves a relationship between two whole numbers. The whole number $A$ is divisible by the natural number $B$ when there is a whole number $Y$ such that $A = B \times Y$. $A$ is divisible by $B$ when the division of $A$ by $B$ results in a remainder of 0. For example, 15 is divisible by 5, but 12 is not divisible by 5, and 0 is divisible by 7 and every other natural number. Other ways of saying that $A$ is divisible by $B$ include saying that $B$ is a factor of $A$ or that $A$ is a multiple of $B$.

It can be misleading to say, "$B$ goes into $A$ evenly." The use of the word *evenly* may evoke the idea that even numbers are involved, although they are not. There is really no reason to use *evenly* when the mathematically precise terms *multiple* and *factor* can be used.

After dividing, people sometimes say, "There is no remainder." That statement is not true. There is always a remainder after division, and those who make this statement probably mean that the remainder is 0. Having a remainder of 0 is not the same as having "no remainder."

In fact, *remainder* and *quotient are* two other important ideas associated with the concept of division. When a whole number is divided by a natural number, it may or may not be that the *dividend* is a multiple of the *divisor*. In general, when a whole number $a$ is divided by a natural number $b$, the result is a number $q$, called the *quotient*, and a whole number $r$, called the *remainder*, such that $a = bq + r$ where $0 \leq r < b$. It should be noted that $q$ and $r$ are the only numbers that satisfy these conditions; that is, they are said to be *unique*. This statement is called the *division algorithm*. For example, the division relationship between 13 and 5 is $13 = (2 \times 5) + 3$. Note that 2 is the quotient, and 3 is the remainder, which is less than the divisor, 5. Bates and Rousseau (1986) provide a number of ways to represent the division algorithm. Reflect 1.8 explores the use of a number line to justify the uniqueness of the quotient and remainder.

See Reflect 1.8 on p. 27.

In many situations, problem solvers need to understand how the divisor, the quotient, and the remainder can be used in the context. Even when they have calculated the quotient and remainder, they need to interpret each of them in the context of the situation and understand how to use each to find an answer. Consider the questions in example 12:

> **Example 12:** Suppose that Tina has 43 oranges that she is packing in into identical empty boxes. Tina experiments and discovers that 8 oranges completely fill a box.

### Reflect 1.8

For each of the following where $a$ is the dividend and $b$ is the divisor find the quotient $q$ and remainder $r$ and write an equation in the form $a = bq + r$.

1. $a = 34$ and $b = 7$

2. $a = 4$ and $b = 9$

3. $a = 20$ and $b = 4$

4. $a = 0$ and $b = 6$

Use a number line to explain how you could find the quotient and remainder in problem 1 above.

How could your number line explanation help someone to understand why there can be only one quotient and only one remainder?

> What is the largest number of boxes that Tina can completely fill?
>
> After completely filling as many boxes as possible, how many "extra" oranges will Tina have?
>
> How many more oranges and boxes will Tina need to pack every orange and completely fill every box with no "extra" oranges?
>
> If Tina packs all the oranges in boxes, what is the smallest number of boxes that she will need? What is the largest number of boxes that she will need?

Observe that the remainder is used in different ways in the different situations that the questions present. Answer these questions, and in particular, carefully explain your interpretations and reasoning for the last two.

Reflect 1.9 presents two examples in which the divisor, 4, takes on different meanings and so is used in different ways. Interpreting what the numbers represent and how they are used is part of developing an understanding of division.

### Reflect 1.9

Create a word problem involving $11 \div 4$ for each of the following situations:

1. The result was 2 groups of 4 with 3 left over.

2. The result was 4 groups of 2 with 3 left over.

Explain how the remainders are being used and the connections of these two examples to the two models for division: *partitive* and *measurement*.

# Properties Underlie Computational Procedures: Big Idea 2

**Big Idea 2.** *The properties of multiplication and addition provide the mathematical foundation for understanding computational procedures for multiplication and division, including mental computation and estimation strategies, invented algorithms, and standard algorithms.*

The computational processes for multiplication and division include both algorithms for finding exact answers and strategies for estimation. Good estimators use a variety of strategies and move flexibly among them (Sowder 1992).

## Using properties for justification and flexibility

**Essential Understanding 2a.** *The commutative and associative properties of multiplication and the distributive property of multiplication over addition ensure flexibility in computations with whole numbers and provide justifications for sequences of computations with them.*

Underlying all of the computational processes for multiplication and division is a deep understanding of place value and the properties of addition and multiplication. These properties include the associative and commutative properties of addition and multiplication and the distributive property of multiplication over addition.

A deep understanding of the operations of multiplication and division requires a deep understanding of place value. For example, being able to decompose a number often helps us in computing. If we are dividing 504 by 8, for instance, thinking of 504 as 48 tens and 24 ones can make the computation easy. Forty-eight tens divided by 8 is 6 tens. Twenty-four divided by 8 is 3. So, the answer is 63.

Because this book focuses on multiplication and division, it does not include a thorough discussion of place value and the properties of addition. Other books in this series discuss the concepts needed for numeration (Dougherty et al. 2010) and the operations of addition and subtraction (Caldwell et al. 2011).

### *Commutative property: $a \times b = b \times a$*

In example 2, discussed previously on pages 12–14, each of Robert's shirts has 4 buttons, and the problem asks how many buttons are on 5 of his shirts. As noted earlier, students might represent the situation by a drawing of 5 shirts, each with 4 buttons. Recall that one interpretation produced the multiplicative expression 5 × 4, and a

For an extended discussion of place value, see *Developing Essential Understanding of Number and Numeration for Teaching Mathematics in Prekindergarten– Grade 2* (Dougherty et al. 2010).

A detailed examination of the properties of addition and subtraction is available in *Developing Essential Understanding of Addition and Subtraction for Teaching Mathematics in Prekindergarten– Grade 2* (Caldwell et al. 2011).

second interpretation yielded the multiplicative expression
4 × 5.

Because the number of buttons is the same in either interpretation, it makes sense to write the equality 5 × 4 = 4 × 5, which is an instance of the commutative property of multiplication. Although 5 × 4 and 4 × 5 have the same numerical value, our interpretations of the two are not the same. With 5 × 4, 5 is the scalar acting on 4, the number of buttons on one shirt. With 4 × 5, 4 is the scalar acting on the number of "first buttons" (or *n*th buttons) on the 5 shirts.

Students quickly realize that the commutative property can make calculations easier, even if they do not know the property by name or use it deliberately. For example, to compute 8 × 2, (8 groups of 2, or 2 + 2 + 2 + 2 + 2 + 2 + 2 + 2), some students sometimes find it easier and more efficient to find 2 × 8 (2 groups of 8 or 8 + 8).

An understanding of the commutative property can make nearly half of the basic multiplication facts easier to learn. In the table in figure 1.12, note that 4 × 2 is directly across the main diagonal from 2 × 4. In fact, because the products *a* × *b* and *b* × *a* are the same for all pairs of numbers *a* and *b*, these products appear in squares that are symmetric with respect to the main diagonal. Because of the commutative property, students learning multiplication facts need to learn only the facts represented by the orange and white cells in the table in figure 1.12.

| × | 0 | 1 | 2 | 3 | 4 | 5 | 6 | 7 | 8 | 9 |
|---|---|---|---|---|---|---|---|---|---|---|
| 0 | 0 | 0 | 0 | 0 | 0 | 0 | 0 | 0 | 0 | 0 |
| 1 | 0 | 1 | 2 | 3 | 4 | 5 | 6 | 7 | 8 | 9 |
| 2 | 0 | 2 | 4 | 6 | 8 | 10 | 12 | 14 | 16 | 18 |
| 3 | 0 | 3 | 6 | 9 | 12 | 15 | 18 | 21 | 24 | 27 |
| 4 | 0 | 4 | 8 | 12 | 16 | 20 | 24 | 28 | 32 | 36 |
| 5 | 0 | 5 | 10 | 15 | 20 | 25 | 30 | 35 | 40 | 45 |
| 6 | 0 | 6 | 12 | 18 | 24 | 30 | 36 | 42 | 48 | 54 |
| 7 | 0 | 7 | 14 | 21 | 28 | 35 | 42 | 49 | 56 | 63 |
| 8 | 0 | 8 | 16 | 24 | 32 | 40 | 48 | 56 | 64 | 72 |
| 9 | 0 | 9 | 18 | 27 | 36 | 45 | 54 | 63 | 72 | 81 |

Fig. 1.12. The symmetry in the multiplication table reflects the commutative property of multiplication for whole numbers.

For division, however, as for subtraction (see Caldwell et al. [2011]), order *does* matter. The following quotients are different:

$$12 \div 3 = 4 \quad \text{and} \quad 3 \div 12 = 1/4$$

One counterexample is sufficient to show that a property does not hold for the system of whole numbers, so we can say that while multiplication and addition are commutative, division and subtraction are not. Reflect 1.10 probes the relationship between the different quotients when the dividend and the divisor are reversed.

---

### Reflect 1.10

Explain how the following two quotients are related:

$$12 \div 3 = 4 \quad \text{and} \quad 3 \div 12 = 1/4$$

Compare this relationship to that between the differences when the minuend and subtrahend are reversed in subtraction:

$$7 - 4 = 3 \quad \text{and} \quad 4 - 7 = -3$$

---

### Associative property: $(a \times b) \times c = a \times (b \times c)$

Multiplicative expressions such as $7 \times 6 \times 5$ are commonplace. However, we can multiply only two numbers at a time—in this case, either 7 and 6 or 6 and 5. So, which one of the two possibilities should we multiply first? We see that we can use parentheses to write $7 \times 6 \times 5$ in two different ways: $(7 \times 6) \times 5$ and $7 \times (6 \times 5)$. Both have the same value, but $7 \times 30$ allows an easier mental calculation than $42 \times 5$. The associative property for multiplication permits us to write the multiplication of three or more whole numbers without using any parentheses and still get the same result, regardless of how we group the factors. For example, what are some different ways to compute $2 \times 3 \times 4 \times 5$ without using the commutative property?

To help understand why the associative property for multiplication holds for whole numbers, we can think about the volume of a rectangular prism with edges of lengths 5, 6, and 7. First, let's consider as the base the face with edges of lengths 6 and 5 (see fig. 1.13a). We can imagine the volume of the prism as composed of layers, each with a "thickness" of 1, stacked on this base. The bottom layer is a $(6 \times 5) \times 1$ rectangular prism with a volume of $6 \times 5$ cubic units. Our prism has 7 layers, resulting in a volume of $7 \times (6 \times 5)$ cubic units. We can turn the prism so that the face with sides of length 6 and 7 is the base (see fig. 1.13b). We now consider the volume as being composed of $1 \times 1 \times 5$ vertical rods, each with a volume of 5 cubic units. We have $7 \times 6$ rods, resulting in a volume of $(7 \times 6) \times 5$ cubic units.

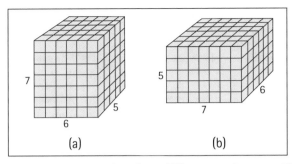

Fig. 1.13. Looking at a prism from different angles to verify the associative property of multiplication of whole numbers

We see that the number of cubic units is the same no matter how we compute the volume, so it makes sense that these two expressions are equivalent—an instance of the associative property for multiplication of whole numbers. We could develop the general statement of the associative property of multiplication for whole numbers in the same way by looking at a rectangular prism with sides of lengths $a$, $b$, and c.

What happens, however, if we consider the idea of associativity with respect to the operation of division? Let's look at an example:

$$(60 \div 6) \div 2 = 10 \div 2 = 5$$
$$60 \div (6 \div 2) = 60 \div 3 = 20$$

These values are different; hence, just as the operation of subtraction is not associative, so also the operation of division is not associative. How to evaluate a numerical expression such as $60 \div 6 \div 2$ is a matter that we will consider later in connection with order of operations.

The power of the commutative and associative properties of multiplication is in the flexibility that they afford in calculations. For example, students may compute the basic fact $4 \times 7$ sometimes by thinking of it as $(2 \times 2) \times 7$ and at other times by thinking of it as $2 \times (2 \times 7)$, options available to them by an application of the associative property. Flexibility in the use of the properties frequently depends on decomposing at least one of the numbers.

Consider the expression $8 \times 17 \times 25$. We can introduce parentheses in either of two places, writing $(8 \times 17) \times 25$ or $8 \times (17 \times 25)$. However, neither grouping provides us with an easy mental calculation. Nevertheless, by using what most upper elementary students know, together with the flexibility that the associative and commutative properties make possible, we can make the computation much simpler. Most students know that four quarters equal a dollar, so they know that $4 \times 25$ is 100. That knowledge makes $8 \times 25$ a value that they can easily obtain mentally, since

$$8 \times 25 = (2 \times 4) \times 25 = 2 \times (4 \times 25) = 2 \times 100 = 200,$$

by an application of the associative property. This process suggests a strategy for mentally calculating $8 \times 17 \times 25$:

| | | |
|---|---|---|
| $(8 \times 17) \times 25$ | $= (17 \times 8) \times 25$ | By the commutative property |
| | $= 17 \times (8 \times 25)$ | By the associative property |
| | $= 17 \times ((2 \times 4) \times 25)$ | Using what we know: $8 = 2 \times 4$ |
| | $= 17 \times (2 \times (4 \times 25))$ | By the associative property |
| | $= 17 \times (2 \times 100)$ | Using what we know: $4 \times 25 = 100$ |
| | $= (17 \times 2) \times 100$ | By the associative property |
| | $= 34 \times 100$ | Using what we know: $17 \times 2 = 34$ |
| | $= 3400$ | Using what we know: $34 \times 100 = 3400$ |

In this instance, we use both the associative and the commutative properties, together with some factual knowledge. Understanding what properties to use and the flexibility that they provide in computational situations are important features of robust understanding. How would you use the commutative and associative properties to simplify the computation of $4 \times 8 \times 15$?

These properties can be extended for the multiplication of more than three numbers. The generalized associative and commutative properties for multiplication mean that factors can be regrouped and rearranged in any order. These properties can be applied to simplify many calculations.

Using the generalized associative property with the expression $5 \times 8 \times 6 \times 4$, we can insert parentheses in different ways and still get the same value, as illustrated below:

$$((5 \times 8) \times 6) \times 4 = (40 \times 6) \times 4 = 240 \times 4 = 960$$
$$5 \times ((8 \times 6) \times 4) = 5 \times (48 \times 4) = 5 \times 192 = 960$$
$$(5 \times 8) \times (6 \times 4) = 40 \times 24 = (10 \times 4) \times 24 = 10 \times (4 \times 24)$$
$$= 10 \times 96 = 960$$

These examples illustrate—but do not prove—that parentheses are not needed for a product with any number of factors. For additional discussion of generalizing and reasoning as we have done with the associative property, see Lannin and colleagues (forthcoming).

For a discussion of mathematical reasoning and generalizing, see *Developing Essential Understanding of Mathematical Reasoning for Teaching Mathematics in Prekindergarten–Grade 8* (Lannin et al. forthcoming).

All of these arrangements, and others that are not shown, have the same value. The generalized commutative and associative properties provide great flexibility in computing the product of three or more factors. Students can choose the arrangement that makes the most sense to them. Consider how they might use both properties to evaluate the expression $5 \times 8 \times 6 \times 4$ in different ways.

The generalized properties tell us that the factors can be rearranged as desired and parentheses can be inserted in any way. By using what they know, students can rearrange and regroup factors to make the calculations simpler. The properties allow them to make the calculations easy for them to perform efficiently. Knowing these properties and how to use them in calculations is part of having a strong understanding of multiplication. An understanding of them is also critical to establishing an understanding of algebra, since they are used extensively in symbolic situations. Understanding these properties is important in creating examples, a task that is explored in Reflect 1.11. Knowledge of these properties also is useful in interpreting and evaluating students' work; see, for example, the various multi-digit multiplication strategies used by Caliandro's (2000) students.

The importance of generalized properties in algebra underlies a big idea in *Developing Essential Understanding of Algebraic Thinking for Teaching Mathematics in Grades 3–5* (Blanton et al. 2011).

### Reflect 1.11

Create at least two examples in which the use of the commutative and associative properties simplifies the calculations.

Commutativity and associativity can also be used effectively in other situations, such as finding the missing factor in $26 \times 42 = \square \times 21$. Recognizing that 21 is half of 42, we can write 42 as $2 \times 21$ and so,

$$26 \times (2 \times 21) = \square \times 21$$
$$(26 \times 2) \times 21 = \square \times 21.$$

Using the associative property to regroup generates the answer of 52. An alternate approach is to reason that because 21 is half of 42, the other quantity must be doubled to maintain equality. Both strategies require the ability to recognize relationships among numbers.

### Distributive property of multiplication over addition: $a \times (b + c) = (a \times b) + (a \times c)$

The power of the distributive property is that it connects the two fundamental operations of addition and multiplication and provides explanations for many additional properties of whole numbers. Our major focus is on the role that this property plays in supporting the development of, and providing explanations for, many of the computational processes associated with multiplication and addition.

Understanding the distributive property is also critical for competency with and understanding of algebraic manipulations.

The usefulness of the distributive property to problem solvers in understanding computational processes is evident in example 13:

> **Example 13:** The fourth-grade class buys 26 fruit baskets to distribute to families at Thanksgiving. Each basket has 18 apples and 23 pears in it. How many pieces of fruit are in the baskets altogether?

Consider the following possible solution strategies:

*First strategy:*     Add to find the number of pieces of fruit in one basket (18 + 23), and then multiply by the number of baskets: 26 (18 + 23).

*Second strategy:*    Multiply to find the total number of apples (26 $\times$ 18) and the total number of pears (26 $\times$ 23), and then add to find the total number of pieces of fruit: (26 $\times$ 18) + (26 $\times$ 23).

Because the number of pieces of fruit is the same, regardless of the strategy, these two expressions will have the same value. The equality, 26 $\times$ (18 + 23) = (26 $\times$ 18) + (26 $\times$ 23), is an instance of the distributive property of multiplication over addition.

An early use of the distributive property is in learning multiplication facts. Students can use the strategy of decomposing one of the factors and then applying facts that they already know to help them find an unknown fact. For example, students can evaluate 6 $\times$ 7 by considering 6 as 3 + 3. Then the problem becomes (3 + 3) $\times$ 7, or the sum of 3 sevens and 3 sevens. That is helpful when 3 $\times$ 7 is a known fact.

Another popular strategy is to decompose one factor into a sum involving a 5. For example, students can evaluate 7 $\times$ 8 by considering 7 as 5 + 2. So, 7 $\times$ 8 = (5 + 2) $\times$ 8. Five eights and 2 eights are 40 + 16, or 56.

An extension of this second strategy is also useful in doing mental calculations. For example, to multiply 5 by 27, one strategy would be to decompose 27 as 20 + 7 and multiply each term by 5, thus obtaining a pair of calculations that are likely to be much easier. This strategy depends not only on the use of the distributive property but also on an understanding of place value. The area representation in figure 1.14 illustrates the reasoning behind this strategy; Englert and Sinicrope (1994) describe the use of such representations with students in grade 4.

The figure shows the partitioning of a 5 $\times$ 27 rectangle into two rectangular parts, one of which is 5 by 20, and the other of which is 5 by 7. The area of the rectangle is 5 $\times$ 27. The area is

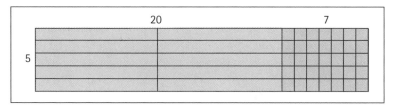

Fig. 1.14. Rectangle representation of 5 × 27

also the sum of 5 × 20 and 5 × 7, or (5 × 20) + (5 × 7). Similar representations can be generalized to illustrate that the distributive property holds for all whole numbers. Reflect 1.12 offers an opportunity to explore the usefulness of the distributive property in mental computations.

> **Reflect 1.12**
>
> Explain how the distributive property might be used to simplify mental computations that can be used to find the following values:
>
> $$23 \times 102 = \square$$
>
> $$14 \times 201 = \square$$

## Identifying properties of division

Essential Understanding 2b. *The right distributive property of division over addition allows computing flexibly and justifying computations with whole numbers, but there is no left distributive property of division over addition and no commutative or associative property of division of whole numbers.*

Is there a distributive property for division over addition? For example, is 60 ÷ (2 + 3) equal to (60 ÷ 2) + (60 ÷ 3)?

$$60 \div (2 + 3) = 60 \div 5 = 12$$
$$(60 \div 2) + (60 \div 3) = 30 + 20 = 50$$

The answers to the calculations are not the same, so we cannot apply distributivity in this division example. However, let's consider a situation in which we write the dividend, rather than the divisor, as a sum. Is (20 + 40) ÷ 5 equal to (20 ÷ 5) + (40 ÷ 5)?

$$(20 + 40) \div 5 = 60 \div 5 = 12$$
$$(20 \div 5) + (40 \div 5) = 4 + 8 = 12$$

This is an instance where one form of the distributive property does work. Mathematicians say that division is *right distributive* over addition but not *left distributive*.

Students can develop both valid and invalid division strategies based on their assumptions about distributivity (Weiland 1985). For some students, incorrect applications of the distributive property can create difficulties in working with algebraic expressions. For example, in simplifying

$$\frac{2x+4}{2},$$

some students will cancel the 2s and get an incorrect answer of $x + 4$, instead of using a correct process:

$$\frac{2x+4}{2} = (2x + 4) \div 2 = (2x \div 2) + (4 \div 2) = x + 2.$$

In a similar fashion, some students will incorrectly obtain

$$\frac{1}{x + 4} \text{ when simplifying } \frac{2}{2x + 4}.$$

This is just one more reason why a strong understanding of the distributive property of multiplication over addition is so important for work with rational numbers. Understanding the properties allows students to reason and make sense rather than simply use procedures memorized by rote or made-up rules that are not correct.

Because $a - b$ also means $a + (-b)$, multiplication also distributes over subtraction. This allows for the use of subtraction in mental calculations of products. For example, to use mental calculation to find the product of 29 and 4, a student might reason as follows: "Thirty fours are just 4 more than 29 fours. So I can multiply 30 and 4 and then subtract 4. That gives 120 – 4, or 116." A symbolic representation of the mental calculations captures the process:

$$29 \times 4 = (30 - 1) \times 4 = (30 \times 4) - (1 \times 4) = 120 - 4 = 116$$

Students can use a similar strategy to compute the product of 17 and 18 mentally. They can think of 18 as 20 – 2, then find the product of 17 and 20, and then subtract $17 \times 2$. The symbolic representation of this mental calculation is equally straightforward:

$$17 \times 18 = 17 \times (20 - 2) = (17 \times 20) - (17 \times 2) = 340 - 34 = 306$$

In each case, the students "change" the problem to use a multiple of 10 because those products are easy to calculate mentally, but they could use any known product. For example, a student who knows that $25 \times 25$ is 625 might multiply $24 \times 25$ by reasoning as follows:

$$24 \times 25 = (25 - 1) \times 25 = (25 \times 25) - (25 \times 1) = 625 - 25 = 600$$

As we discussed above, the distributive property can also be used with division when the dividend can be decomposed. For example, a student dividing 6003 by 3 could decompose 6003 into $6000 + 3$. Then the mental calculation would become quite simple, as its symbolic representation shows:

$$6003 \div 3 = (6000 + 3) \div 3 = (6000 \div 3) + (3 \div 3) = 2000 + 1 = 2001$$

Another student mentally dividing 344 by 8 could proceed by using two known facts: $320 \div 8 = 40$, and $24 \div 8 = 3$. A symbolic representation of the process follows:

$$344 \div 8 = (320 + 24) \div 8 = (320 \div 8) + (24 \div 8) = 40 + 3 = 43$$

You may find it useful to examine the connections between the arithmetic examples above and the algebraic examples immediately preceding them. Note also that the strategy of decomposing the divisor usually leads to an incorrect answer. For example, the quotient $100 \div 20$ is not the same as the sum of quotients $(100 \div 10) + (100 \div 10)$.

The distributive property can also help in comparing products. Which is greater, for example, $35 \times 46$ or $36 \times 45$? We can think of $35 \times 46$ as 35 groups of 46 objects. Decomposing 46 as $45 + 1$, we can think of the product as 35 groups of 45 objects and another 35 groups each with one object. We can illustrate this situation with an area representation showing rectangular regions, as in figure 1.15a.

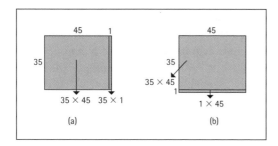

Fig. 1.15. Rectangles representing the decompositions of $35 \times 46$ and $36 \times 45$

Similarly, we can think of $36 \times 45$ as 36 groups of 45 items, or as 35 groups of 45 items and another group of 45 items. A pictorial representation appears in figure 1.15b. Each of the diagrams shows parts labeled $35 \times 45$, which are the same. In the first case, however, there are 35 "extras," whereas in the second case, there are 45 "extras." The following symbolic expressions represent the decomposition process used to obtain the products:

$$35 \times 46 = 35 \times (45 + 1) = (35 \times 45) + (35 \times 1) = 35 \times 45 + 35$$
$$36 \times 45 = (35 + 1) \times 45 = (35 \times 45) + (1 \times 45) = 35 \times 45 + 45$$

So, $36 \times 45$ is greater than $35 \times 46$. Notice the similarities and differences between this comparison of $35 \times 46$ and $36 \times 45$ and the comparison in Reflect 1.13.

The distributive property is also helpful in estimation, as example 14 illustrates:

**Example 14:** Estimate the unit cost in cents of an 8-ounce jar of olives costing $5.93.

See Reflect 1.13 on p. 38.

**Reflect 1.13**

Use the distributive property to compare 43 × 57 and 47 × 53.

Which is greater? By how much?

Draw diagrams and explain.

One strategy that we might use to make the estimate would be to choose a multiple of 8 by using our knowledge of the basic facts. Because 8 × 7 is 56, 8 × 70 is 560, and $5.60 is close to $5.93. But 70 cents is clearly an underestimate. We can find a more precise estimate by using the right distributive property of division over addition. We can decompose 593 into a sum of an easily identifiable multiple of 8 and some other number and then replace the other number with a multiple of 8 that is close to the other number. For example, 593 = 560 + 33. Since 32 is a multiple of 8 that is close to 33, 593 ÷ 8 is about the same as

$$(560 + 32) \div 8 = (560 \div 8) + (32 \div 8) = 70 + 4 = 74.$$

The distributive property of multiplication over addition involves a sum of only two terms. However, a generalization of the property extends to the distribution of multiplication over more than two terms. For example,

$$6 \times (3 + 4 + 5) = (6 \times 3) + (6 \times 4) + (6 \times 5).$$

We can provide a justification for this result by using the distributive property of multiplication over addition twice:

$$6 \times (3 + 4 + 5) = 6 \times (3 + (4 + 5)) = (6 \times 3) + 6 \times (4 + 5) =$$
$$(6 \times 3) + (6 \times 4) + (6 \times 5)$$

The generalized distributive property of multiplication over addition for several terms is justified by the repeated use of the distributive property of multiplication over addition.

## Avoiding ambiguity through order of operations

Essential Understanding 2c. Order of operations *is a set of conventions that eliminates ambiguity in, and multiple values for, numerical expressions involving multiple operations.*

If the symbolic expression that we write to solve a problem involves more than one operation, which operation should we perform first? This is the important question that example 15 introduces:

**Example 15:** Susie and Harry each entered 3 + 4 × 5 into their calculators. Susie got an answer of 23. Harry got an answer of 35. Are they both right? How could they interpret their answers?

Susie and Harry produced two different answers: 23 and 35. Before exploring other aspects of the situation, consider Reflect 1.14 to examine how the students might have arrived at their answers.

### Reflect 1.14

What operations did Susie's calculator use to get 23?

What operations did Harry's calculator use to get 35?

Should this type of expression have two correct answers? Obviously, no—we want only one answer. When scientists are using an expression to determine the location of a satellite, for example, it is critical that all who are working with that expression evaluate it in exactly the same way. To avoid ambiguity, mathematicians have agreed on a set of rules to follow in performing calculations. This set of conventions ensures that there is no confusion about how to interpret the order in which to perform operations. These are not properties, nor are they theorems or axioms. The rules are as follows:

1. Perform operations inside **parentheses**.

2. Perform operations with **exponents**.

3. Proceeding from left to right, perform all **multiplication and division**.

4. Again proceeding from left to right, perform all **addition and subtraction**.

This convention is frequently called *order of operations*. Susie's calculator is following these guidelines when it first multiplies 4 and 5 and then adds 3 to get an answer of 23. Harry's calculator, by contrast, is performing the operations in a strictly left-to-right order. This is common on some calculators with no parentheses; once two numbers are entered with an operation, the calculator must perform the operation to be able to proceed. Harry's calculator added 3 and 4 and then multiplied the result by 5. This answer is not consistent with the guidelines, because it does not follow the order of operations. It would be an incorrect answer in a context with an expectation that the order of operations would be followed. Reflect 1.15 revisits familiar ideas related to order of operations and other properties and meanings of multiplication.

See Reflect 1.15 on p. 40.

Performing operations in the correct order is especially important in algebra. For example, it is common to simplify an expression such as $7 - 5(2x - 3)$. Properly manipulating this expression depends on understanding the order of operations. For students to develop both proficiency with and understanding of the order of operations, they need extensive early experiences with expressions that involve multiplication, division, addition, subtraction, and parentheses.

## Using properties to justify algorithms

*Essential Understanding 2d. Properties of whole numbers justify written and mental computational algorithms, standard and invented.*

Proficiency with the use of written and mental computational algorithms, standard or nonstandard, contributes to a robust understanding of multiplication and division. Understanding of these operations is limited if it does not also include an understanding of how and why these algorithms work. Even in the simplest of computations, what properties provide a justification is not always obvious. Can we provide a justification for the sum 30 + 40 = 70 other than a direct modeling, counting on from the original set of 30 objects until we have added 40 more objects? The following process might show a more insightful understanding: we could rewrite 30 as 3 × 10 and 40 as 4 × 10 by using our knowledge of place value and writing

$$30 + 40 = (3 \times 10) + (4 \times 10) = (3 + 4) \times 10 = 7 \times 10 = 70.$$

Note the powerful use of the distributive property. Three groups of 10 and 4 groups of 10 are 7 groups of 10, but this explanation still makes implicit use of the distributive property.

### Justifying algorithms for multiplication

The multiplication of 30 and 40 involves even more use of the properties. We can begin by rewriting 30 and 40 as before:

$$30 \times 40 = (3 \times 10) \times (4 \times 10)$$

Using the generalized associative and commutative properties, we can rewrite this as

$$(3 \times 4) \times (10 \times 10) = 12 \times 100.$$

Using multiplication facts (3 × 4 = 12) and place value, we can rewrite this as (10 + 2) $\times$ 100. With the distributive and commutative properties, we have 10 $\times$ 100 + 2 $\times$ 100, an expression that our knowledge of place value allows us to write as 1200. Explanations for even the most obvious computations rely heavily on the properties of addition and multiplication. It is important that we be aware of our use of these properties when we perform multiplicative computations. This knowledge of the properties helps us to develop a deeper understanding of the algorithms.

Previously, we looked at the multiplication of 5 and 27 in terms of place value and the distributive property of multiplication over addition (see fig. 1.14 and the accompanying discussion). Now let's consider a similar example, 4 $\times$ 23, in even closer detail. We know that we can represent this multiplication by means of a rectangle with side lengths 4 and 23, divided into parts as illustrated in figure 1.16. The area of one shaded part is 80, and the area of the other shaded part is 12, so the product of 4 and 23 is the sum of the shaded areas, or 92.

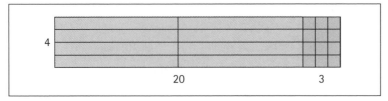

Fig. 1.16. An area model for 4 $\times$ 23

We can also evaluate this product by using the distributive property. Let's do so very deliberately, giving an explanation for each step:

| 4 $\times$ 23 | = 4 $\times$ (20 + 3) | What we know about place value |
| | = (4 $\times$ 20) + (4 $\times$ 3) | Distributive property |
| | = 80 + 12 | Multiplication and place value |
| | = 80 + (10 + 2) | What we know about place value |
| | = (80 + 10) + 2 | Associative property for addition |
| | = 90 + 2 | Addition and place value |
| | = 92 | Place value |

Reflect 1.16 invites you to give a similar explanation for a different product.

In line 3 of our calculations above for 4 $\times$ 23, note that the numbers 80 and 12 are the result of multiplying 20 and 4, and 3 and 4, respectively. Also note the connection between line 1 and

stop
reflect

See Reflect 1.16 on p. 42.

Reflect 1.16

Use and state the properties of multiplication and addition, as well as the concept of place value, to justify the steps in multiplying 38 and 7.

the area representation for the multiplication. Frequently, these observations are combined into two different algorithms. One of these processes, shown in figure 1.17a, could describe a mental calculation: we could think of 23 as 20 + 3, then multiply each term by 4, and then add the two products. The other process, illustrated in figure 1.17b, is similar to a standard algorithm.

|  |  |  |  |
|---|---|---|---|
| 23 |  | 23 |  |
| $\times\ 4$ |  | $\times\ 4$ |  |
| 80 | $4 \times 20$ | 12 | $4 \times 3$ |
| 12 | $4 \times 3$ | 80 | $4 \times 20$ |
| 92 |  | 92 |  |
| (a) |  | (b) |  |

Fig. 1.17. Two multiplication algorithms

The numbers 12 and 80 are called *partial products*, and each of the processes illustrated in the figure is sometimes called a *partial products algorithm*. Notice how and where the partial products also appear in the area representation. We can see the fundamental role that the concept of place value and knowledge of the properties—especially the distributive property—play in finding the product of 4 and 23.

Let's look carefully at how we write 23 $\times$ 4 when we use an algorithm that is more standard. The format might be something like that shown in Reflect 1.17, where the algorithm is presented for examination.

Consider a more complex problem: multiplying 24 and 35. Suppose that we represent the multiplication by a rectangle with side lengths 24 and 35, like that shown in figure 1.18. We can partition the side with length 35 into 35 equal segments, creating 35 smaller rectangles, each 1 $\times$ 24, as illustrated in the figure. We can then group these smaller rectangles into four larger rectangles with 3 groups of 10 smaller rectangles and 1 group of 5 smaller rectangles. Then we can partition the side of length 24 into two segments, each of length 10, and 4 segments, each of length 1. Although it is not necessary to use partitions that are multiples of 10, they allow us to make connections with other computational algorithms more readily. Understanding place value is critical to understanding these ideas and connections.

See Reflect 1.17 on p. 43.

## Reflect 1.17

Examine the written form of a more standard algorithm for the product of 23 and 4:

```
   1
  23
× 4
────
  92
```

1. Why is 2 written in the answer line? What does it mean?

2. What does the "carried" 1 mean? Why is it placed where it is? What other term(s) besides *carry* could be used to develop a more meaningful understanding?

3. Why is 9 placed where it is? What does it represent?

4. How would your answers to questions 1–3 help provide a mathematical justification for a traditional, standard algorithm? How would they contribute to a deeper understanding of the traditional algorithm?

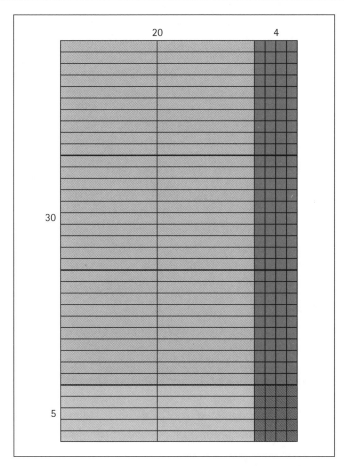

Fig. 1.18. An area model for 35 × 24

An examination of the area representation reveals that the four rectangles that appear in different shades of color have areas of 20, 100, 120, and 600, for a total area of 840, and 840 is then the product of 24 and 35.

The calculation of the product, as justified by the properties used, is as follows:

$24 \times 35 = 24 \times (30 + 5)$                                     Place value

$= (24 \times 30) + (24 \times 5)$                                          Distributive property

$= ((20 + 4) \times 30) + ((20 + 4) \times 5)$    Place value

$= (20 \times 30) + (4 \times 30) +$                             Distributive property
$(20 \times 5) + (4 \times 5)$

$= 600 + 120 + 100 + 20$                                       Multiplication and place value

$= 840$                                                                              ???

Justifying the last line involves the kind of reasoning that we have used before. Can you provide the justification? Also note how the numbers 20, 100, 120, and 600 appear in the steps. These calculations depend on understanding place value—specifically, the knowledge that 24 is 20 + 4 and that 35 is 30 + 5—and the use of the distributive property.

These observations give rise to another partial products algorithm, which in the case of 24 × 35 has the following written form:

$$
\begin{array}{r}
24 \\
\times\, 35 \\
\hline
20 \\
100 \\
120 \\
600 \\
\hline
840
\end{array}
$$

What is the source of the numbers 20, 100, 120, and 600 in this algorithm? These calculations provide an instance of a very important procedure in algebra—the multiplication of binomials:

$$(a + b) \times (c + d) = a \times c + a \times d + b \times c + b \times d$$

Students should recognize this as the repeated use of the distributive property. Reflect 1.18 invites consideration of the more traditional algorithm.

These same ideas can be used to develop and extend the computations of products of multi-digit numbers with any number of places. The preceding examples suggest a framework for developing algorithms for multiplying multi-digit numbers. To explore a multi-digit example, find the product of 246 and 357 by using a partial product algorithm and a more traditional algorithm and examine the connections between the two algorithms.

See Reflect 1.18 on p. 45.

## Reflect 1.18

Examine the use of the traditional algorithm in the case of 24 × 35:

$$
\begin{array}{r}
1,2 \\
24 \\
\times\ 35 \\
\hline
120 \\
72\phantom{0} \\
\hline
840
\end{array}
$$

1. What does the "carried" 2 mean? Why is it written where it is? Can you think of another term besides *carry* that might help to develop a more meaningful understanding?

2. Why is 120 written below the first horizontal ("total") line? What does it represent?

3. Why is a 1 "carried"? What does it represent? Why is it written where it is? Again, what term might be a better choice than *carry*?

4. What does the 72 represent? Why?

5. How are these numbers connected to the numbers in the area representation and to the previous partial product algorithm?

### *Justifying algorithms for division*

The long division algorithm is often difficult for students to use and understand. However, when teachers present an abbreviated form, students' understanding is often sacrificed. Students demonstrate less proficiency in carrying out the algorithm and make more errors.

Everyday contexts present many long division situations. Consider the problem in example 16:

> **Example 16:** Jason and a dozen of his friends saw an advertisement for an online purchase of music that they could download and share. They read the conditions of the purchase, and they saw that the 13 of them would have to buy 4230 tunes over a period of a year and pay in advance. When they realized how many tunes each of them would have to buy if they all bought equal numbers of tunes, they decided it was not such a good deal after all. How many tunes would each of them have to buy?

This type of situation involves sharing (recall example 5 with Grandfather and his coins).

We can solve this problem by dividing 4230 by 13, using the long division algorithm in a format like that shown in figure 1.19. Explaining the meanings of the numbers requires a strong understanding of division. Reflect 1.19 examines the steps in the process and probes the numbers' meanings.

$$
\begin{array}{r}
3 \\
13\overline{)4230} \\
39 \\
\hline
3
\end{array}
\qquad
\begin{array}{r}
32 \\
13\overline{)4230} \\
39 \\
\hline
33 \\
26 \\
\hline
7
\end{array}
\qquad
\begin{array}{r}
325 \\
13\overline{)4230} \\
39 \\
\hline
33 \\
26 \\
\hline
70 \\
65 \\
\hline
5
\end{array}
$$

Fig. 1.19. The long division algorithm for dividing 4230 by 13

## Reflect 1.19

Consider the context of example 16 while examining the format of the long division algorithm shown in figure 1.19 for 4230 ÷ 13.

1. What does the 3 in the quotient represent? Why is it needed?

2. What does the 39 represent?

3. Why subtract?

4. What does the 3 after this subtraction represent?

5. What does the 33 represent? Why is it needed?

6. In explaining this algorithm to someone, people commonly ask a question like, "How many times does 13 go into 42?" What does this question mean? Is there a better way of formulating it mathematically?

7. Write an equation that connects the dividend, the quotient, the divisor, and the remainder.

Example 17 presents a different everyday long division situation that we can use to examine some other strategies for solving a division problem:

**Example 17:** Again, Jason and his dozen friends are looking online for great deals, and this time they see an ad for the latest cell phone, which the manufacturer claims has many marvelous features. Jason takes the initiative and orders 13 phones. When the phones arrive, they unfortunately do not work as advertised, and Jason ends up returning them. He gets back a refund check of $4225, which he cashes at his bank so that he can repay his friends. How much cash will Jason and each friend receive from the refund? Will they have any money left over?

This situation again involves sharing fairly, so the answer can be obtained by dividing 4225 by 13 and finding the quotient and remainder. The discussion that follows describes various strategies for distributing the refund and shows corresponding symbolic representations. Take time on your own to look for and explain connections between these strategies and the traditional long division algorithm presented above.

When Jason goes to the bank to obtain $4225 in cash to repay his friends and himself, what denominations of bills should he get, and how many bills should he get of each denomination? His choices range from 4225 ones to 4 one thousand dollar bills, 2 one hundred dollar bills, 2 tens, and 1 five. What denominations to get and how many of each are choices that will be influenced by the strategy that Jason chooses to use to distribute the money.

One strategy would be for Jason to get 422 tens and 1 five and distribute 1 ten to each of his 12 friends while keeping 1 ten for himself. This first step of the distribution would leave Jason with 409 tens and 1 five after giving each friend and himself ten dollars apiece. We could represent the number of dollars remaining symbolically as

$$4225 - 13(10) = (422(10) + 5) - 13(10) = 409(10) + 5.$$

In step 2, Jason could repeat the process from step 1, leaving 396 tens and 1 five, with each friend and Jason now having 20 dollars. We could now write the number of remaining dollars as

$$
\begin{aligned}
4225 - 13(10) - 13(10) &= (422(10) + 5) - 13(10) - 13(10) \\
&= 409(10) + 5 - 13(10) \\
&= 396(10) + 5.
\end{aligned}
$$

Although Jason could continue this distribution strategy and find an answer, the strategy would be inefficient. Using base-ten blocks to model the situation would quickly show its inefficiency.

A second strategy that Jason might use would be to get 42 one hundred dollar bills, 2 tens, and 1 five and give 1 one hundred dollar bill to each of his friends while keeping 1 one hundred dollar bill for himself. This first step would leave 29 one hundred dollar bills, 2 tens, and 1 five, after giving each friend and Jason a hundred dollars apiece. We could represent the number of remaining dollars as

$$4225 - 13(100) = (42(100) + 2(10) + 5) - 13(100) = 29(100) + 2(10) + 5.$$

Again, the second step could repeat the process, leaving 16 one hundred dollar bills, 2 tens, and 1 five, with each friend and Jason now having 200 dollars apiece. Step 3, repeating the same process, would leave 3 one hundred dollar bills, 2 tens, and 1 five. At this point, Jason could no longer continue giving one hundred dollar bills, because he would not have enough for each of his 12 friends and himself. Now Jason might go to the bank and request 30 tens in exchange for the remaining 3 one hundred bills, giving him a total of 32 tens. He could then distribute the tens one at a time as before, with each friend and Jason himself now getting 2 tens, leaving 6 tens. At this point, Jason would again need to go to the bank, this time to exchange the 6 tens for 12 fives, giving him a total of 13 fives. So now each friend would get 1 five, as would Jason. Altogether, each friend and Jason each would get 3 one hundred dollar bills, 2 tens, and 1 five, for a total of 325 dollars. Reflect 1.20 explores possibilities for representing this strategy in different ways.

### Reflect 1.20

Think about Jason's second strategy and how it can be modeled.

1. Explain how base-ten blocks could be used to model this strategy, paying particular attention to when Jason would run out of a denomination of bills.

2. Give a symbolic representation for this strategy.

3. Describe connections between this strategy and the long division algorithm presented for example 16 .

Figure 1.20a shows a different symbolic representation of Jason's second strategy. However, Jason should realize that he would not need to distribute the one hundred dollar bills and ten dollar bills one a time. Instead he could distribute larger multiples of these bills. A symbolic representation of this strategy appears in figure 1.20b. Carefully explore the connections between the two symbolic representations.

The preceding examples have shown *partitive* division, which involves fair sharing. Example 18 shows a different kind of situation:

> **Example 18:** Jason and his classmates want to distribute food baskets to as many families as possible at Thanksgiving. By working hard, they have been able to purchase 4230 food items, and they plan to put 13 items in each basket. To how many families can they give a food basket?

This example presents a type of division problem that we call a *measurement* problem (recall example 4, with Grandmother giving out 7 marbles to each of her grandchildren). Dividing 4230 by 13 will provide the answer, and the long division algorithm would have a format similar to that used with the sharing situation in example 16. Understanding division allows us to explain the numbers and operations in the representation. Reflect 1.21 seeks such explanations in the context of the measurement situation in example 18.

See Reflect 1.21
on p. 50.

```
              5
             10
             10
            100
            100                         5
            100                        20
      13) 4225                        300
          1300                  13) 4225
          2925                      3900
          1300                       325
          1625                       260
          1300                        65
           325                        65
           130                         0
           195
           130
            65
            65
             0
           (a)                        (b)
```

Fig. 1.20. Symbolic representations of Jason's second strategy,
in (a) the original and (b) a more efficient form

## Reflect 1.21

In example 18, Jason and his classmates have 4230 food items and plan to
make food baskets with 13 items in each. The long division algorithm for this
measurement situation would have a format like that for the sharing problem
in example 16:

```
          325
    13) 4230
        39
        33
        26
        70
        65
         5
```

1. What does the 3 in the quotient represent?

2. What does the 39 represent?

3. What does the 3 after the subtraction represent?

4. What does the 33 represent?

5. Ask and answer questions similar to the previous ones about the rest of the
   algorithm.

6. Would Jason and his classmates have food items left over?

# Conclusion

This chapter has provided a framework on which teachers of grades 3–5 can build a deeper understanding of multiplication and division. Describing multiplication as a scalar operation embraces the many representations and problem situations for which the operation of multiplication is an appropriate choice. Recognizing the many ways to reason multiplicatively and the connections among them is equally important.

The chapter has also examined the properties of multiplication and division more deeply to make sense of the foundations for computational procedures, standard and nonstandard, written and mental. Our deeper understanding of procedures gives us a better understanding of our students' thinking about their work, as chapter 3 will show. A deep understanding of multiplication prepares us in turn to guide our students to—

- perform computations accurately; and

- comprehend and explain the computational procedures.

Such abilities allow students to generalize computational procedures, use them flexibly and fluently, and consider them as valuable assets in understanding multiplication and division.

# Chapter 2

# Connections: Looking Back and Ahead in Learning

Multiplication and division are powerful ideas. Their use in computations and problem solving in grades 3–5 is only part of their contribution to the curriculum and to students' knowledge. This chapter examines connections between the essential understandings described in chapter 1 and other mathematical topics, while considering the foundations of these ideas in the primary grades and their application in subsequent grades.

An understanding of the operation of multiplication builds on ideas about place value, addition, and skip counting. It aids in developing an understanding of fractions and percents and lays a foundation for future work with ratios and similarity. An understanding of algorithms and the use of properties of whole number multiplication and division to justify them provides a basis for developing fluency in rational number computations and algebraic manipulations.

## Multiplication as a Fundamental Operation

Each of the mathematical topics in this chapter connects with the big ideas and essential understandings described in chapter 1. The scalar definition of multiplication (Essential Understanding 1a) and the properties of multiplication are evident in other mathematical topics that students encounter throughout their school years. From place value, addition, and skip counting to percent, ratio, and similarity, thinking about the fundamental operation of multiplication as a scalar process connects concepts, procedures, and problems.

Essential
Understanding 1a

*In the multiplicative expression $A \times B$, $A$ can be defined as a scaling factor.*

### Place value and addition

The base-ten numeration system is fundamental to understanding numbers and any written or mental algorithm with whole numbers.

These are key topics in the primary grades, where students learn that 23 is 2 tens and 3 ones, which they later write in expanded notation as $2 \times 10 + 3 \times 1$. In this expression, 2 is the scaling factor that tells how many tens, and 3 is the scaling factor that tells how many ones. When students learn to add 23 and 35, they add each place value separately. The connections of multiplication to place value are integrated into the addition algorithm through the use of the distributive property, together with the commutative and associative properties for addition:

$$23 + 35 = (2 \times 10 + 3) + (3 \times 10 + 5) \qquad \text{Place value}$$
$$= (2 \times 10 + 3 \times 10) + (3 + 5) \qquad \text{Addition properties}$$
$$= 5 \times 10 + 8 \qquad \text{Distributive property}$$
$$= 58 \qquad \text{Place value}$$

Understanding the interplay of place value and the operations of addition and multiplication is a major part of developing a robust understanding of whole number operations and is critical to decomposing factors and combining partial products.

## Division and subtraction

Division is defined in terms of multiplication in much the same way that subtraction is defined in terms of addition. In fact, the symbolic descriptions of subtraction and division are quite similar. We recall that $a - b = a + (-b)$, where $-b$ is the notation for the additive inverse (or opposite) of $b$, and $a \div b = a \times (1/b)$, where $1/b$ is the notation for the multiplicative inverse (or reciprocal) of $b$, with $b \neq 0$. Of course, $b$ cannot be 0, because such a case would involve division by zero, which is undefined.

Alternate and equivalent descriptions for subtraction and division are, respectively, that $a - b$ is defined to be the number $y$ when $a = b + y$ and that $a \div b$ is defined to be the number $y$ when $a = b \times y$, with $b \neq 0$. The similarity in symbols reflects the similarities in defining subtraction and division in terms of addition and subtraction, respectively.

## Skip counting

Skip counting is a worthwhile activity that many primary teachers use with their students, and it is often viewed as an example of using the operation of addition. However, it is also multiplicative in nature. Furthermore, many other important mathematical ideas are embedded in skip counting.

Consider skip counting by 6: 6, 12, 18, 24, and so forth. These numbers form a sequence, and each natural number corresponds to a term in the sequence. The first term is $1 \times 6$, the second term is 2 × 6, the third term is 3 × 6, and so on. The position of the term in the sequence multiplied by 6 is that term in the sequence, as shown in figure 2.1.

| Position in the sequence | 1 | 2 | 3 | 4 | 5 | 6 | 7 |
|---|---|---|---|---|---|---|---|
| Term in the sequence | 6 | 12 | 18 | 24 | 30 | 36 | 42 |

Fig. 2.1. A sequence created by skip counting by 6, starting at 6

What is the fifteenth term in the sequence? Understanding multiplication as a scalar operation helps us to produce the answer to this question (Essential Understanding 1d) when we observe that the number of the term in the sequence acts like a scalar operating on 6.

Working in this way with a sequence produced by skip counting also provides an early opportunity for algebraic reasoning. For any natural number $n$, what algebraic expression can we use to represent the $n$th term in the sequence? Every term in the sequence has a factor of 6 or, equivalently, is divisible by 6. Conversely, every natural number that is a multiple of 6 is in the sequence. If $t$ is a number in the sequence, what is the position number of $t$?

This type of sequence is called *arithmetic* because the difference between any two consecutive terms is constant—in this case, 6. Skip counting also describes a *function*, with an *input* value and a unique, corresponding *output* value. The input value is a natural number and the output value is both a multiple of the input value and, in this case, a multiple of 6. We can represent this function algebraically by writing $f(x) = 6x$. A graph of this function is shown in figure 2.2. Cooney, Beckmann, and Lloyd (2010) elaborate on the connection between sequences and functions.

Essential Understanding 1d

*A scalar definition of multiplication is useful in representing and solving problems beyond whole number multiplication and division.*

*Developing Essential Understanding of Functions for Teaching Mathematics in Grades 9–12* (Cooney, Beckmann, and Lloyd 2010) presents a full discussion of connections between sequences and functions.

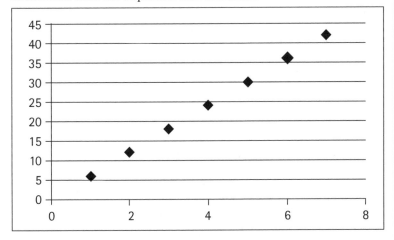

Fig. 2.2. A graph representing the function created by skip counting by 6, starting at 6

Both the table in figure 2.1 and the graph of the skip counting function in figure 2.2 show relationships between changes in the position numbers and values of the corresponding terms. Examination of these relationships highlights two very important ideas.

The first is the idea of a proportional relationship that can be found in some skip counting examples. If a position number for the sequence 6, 12, 18, 24, ... is doubled, then the value of the term in the sequence is also doubled. In general, for this sequence, multiplying a position number by a natural number $p$ also multiplies the value of the term in the sequence by $p$. For example, multiplying the second position number, 2, by 3 yields the result 6, or the sixth position number. The corresponding sequence term for the second position number is 12, which when multiplied by 3 is 36, the corresponding sequence term for the sixth position number. Recognizing these connections involves the use of proportional reasoning. Lobato and Ellis (2010) provide further discussion of how proportional reasoning relates to functions and contexts.

The second idea that emerges from examining the table and the graph is the idea of an additive interpretation of the rate of change. If the sequence position number is increased by 1, then the value of the term in the sequence is increased by 6, and if the increase in the position number is 2, then the value of the term is increased by 12. An additive change in the position number results in a corresponding additive change in the value of the sequence term. From an algebraic point of view, this is an example of the concept of *slope*, a rate of change.

If the skip counting by 6 starts at 4, we then have the sequence 4, 10, 16, 22, and so forth, as shown in the table in figure 2.3. Again, we have an arithmetic sequence with a constant difference of 6 between any two consecutive terms. Is this sequence a function? Why? How could we represent it algebraically? What would its graph look like?

| Position in the sequence | 1 | 2 | 3 | 4 | 5 | 6 | 7 |
|---|---|---|---|---|---|---|---|
| Term in the sequence | 4 | 10 | 16 | 22 | 28 | 34 | 40 |

Fig. 2.3. A sequence created by skip counting by 6, beginning at 4

An examination of the table in the figure shows that we cannot use proportional reasoning to relate the position numbers and terms of this sequence. For example, if we double the position number 3 for a position number of 6, doubling the corresponding term in the sequence, 16, does not give us the term that corresponds to the new position number. However, our scrutiny of the table also shows that we can still use this example of skip counting to illustrate the concept of slope. The sequence has a constant rate of change. Another observation that we can make is that in the first example of skip counting, all the terms in the sequence are divisible by 6, and in the

*Developing Essential Understanding of Ratios, Proportions, and Proportional Reasoning for Teaching Mathematics in Grades 6–8 (Lobato and Ellis 2010) presents a full discussion of proportionality in mathematics and everyday life.*

second example, when we divide the terms by 6, we always get a remainder of 4.

If we skip count by 6 starting at 0, then we have the sequence 0, 6, 12, 18, 24, and so forth. How would tables and graphs for skip counting by 6 starting at 0 and starting at 6 compare? How are they similar? How are they different?

Finally, we can use skip counting to introduce students to the idea of an infinite sequence because the values of the terms in the sequence can continue indefinitely. This example of an infinite sequence establishes a one-to-one correspondence between an infinite set and a proper subset of it, the counting numbers, a common definition of an infinite set.

## Divisibility and prime factorization

One way to determine if a number is divisible by another number without carrying out a division algorithm is to find the prime factorization of that number. The justification for this strategy relies on the commutative and associative properties of multiplication (Essential Understanding 2a). The *fundamental theorem of arithmetic*—also known as the *unique factorization theorem for arithmetic*—states that every natural number except 1 can be represented as a product of one or more prime numbers in exactly one way, without regard to the order in which the primes are written. The resulting product is the prime factorization of the number. For example, the prime factorization of 2520 is

$$2 \times 2 \times 2 \times 3 \times 3 \times 5 \times 7.$$

A prime number has exactly two different factors, the number itself and 1.

Chapter 1 briefly explored the use of a tree diagram to model a multiplicative situation such as that in example 6, about the different outfits that Sue can make with 3 shirts and 2 skirts. A tree diagram can also be useful in finding the prime factors of a whole number. For example, figure 2.4 shows a tree diagram used as a solution strategy for finding the prime factorization of 2520.

The prime factorization of a natural number can be used to determine what numbers are multiples or divisors of that number. For example, we can see that $2 \times 3 \times 7$ is a divisor of 2520, since each factor appears in the prime factorization. However, $3 \times 5 \times 11$ is not a divisor of 2520, since 11 is not a factor of 2520. By multiplying 2520 by $2 \times 2$, we see that

$$\underbrace{2 \times 2 \times 2 \times 2 \times 2 \times 3 \times 3 \times 5 \times 7}_{2520}$$

is a multiple of 2520.

Essential ← Understanding 2a

*The commutative and associative properties of multiplication and the distributive property of multiplication over addition ensure flexibility in computations with whole numbers and provide justifications for sequences of computations with them.*

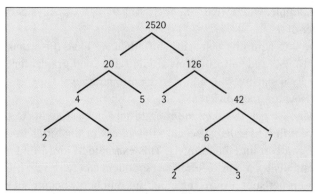

Fig. 2.4. Using a tree diagram to find the prime factorization of 2520:
2 × 2 × 2 × 3 × 3 × 5 × 7

We can find the *greatest common factor* and *least common multiple* of two natural numbers by using their prime factorizations. Consider the numbers 2520 and 9900:

$$2520 = 2 \times 2 \times 2 \times 3 \times 3 \times 5 \times 7$$
$$9900 = 2 \times 2 \times 3 \times 3 \times 5 \times 5 \times 11$$

Their common factors, shown above in orange, indicate that the greatest common factor is $2 \times 2 \times 3 \times 3 \times 5$. Their least common multiple includes not only the common factors but also each of the other prime factors: $2 \times 2 \times 2 \times 3 \times 3 \times 5 \times 5 \times 7 \times 11$. Notice how we can use the commutative and associative properties of multiplication of whole numbers to justify our claims about the prime factorizations.

We have seen how the prime factorization of a natural number can assist us in determining whether that number is divisible by another without doing the actual division. We can do this in other ways as well. For example, a number is divisible by 10 if the last digit is 0. To justify a case of this claim, suppose that *ABC* is a three-digit number, and the last digit, *C*, is 0. Then, using the expanded form, we can rewrite *ABC* as $A \times 10^2 + B \times 10^1 + C \times 10^0$. However, since $C = 0$, we can also write *ABC* in the form $A \times 10^2 + B \times 10^1$. The use of the distributive property gives us $ABC = 10 \times (A \times 10^1 + B \times 10^0)$, and this means that 10 is a divisor of *ABC*.

Many other "rules," or tests, describe conditions under which whole numbers are divisible by another number. The most common of these describe conditions under which numbers are divisible by 2, 3, or 5. There are also tests for divisibility by 4, 6, 8, and 9. Can you find these conditions for divisibility?

Another way to look at divisibility focuses on remainders. Suppose that we arrange the whole numbers from 0 to 41 sequentially in a table with 6 columns, as shown in figure 2.5. Note that the difference of any two numbers in the same column is a multiple of 6, and that when we divide any numbers in the same column by

6, we get the same remainder. Also note that the entries in the first row of the table tell us what that remainder is.

| 0 | 1 | 2 | 3 | 4 | 5 |
|---|---|---|---|---|---|
| 6 | 7 | 8 | 9 | 10 | 11 |
| 12 | 13 | 14 | 15 | 16 | 17 |
| 18 | 19 | 20 | 21 | 22 | 23 |
| 24 | 25 | 26 | 27 | 28 | 29 |
| 30 | 31 | 32 | 33 | 34 | 35 |
| 36 | 37 | 38 | 39 | 40 | 41 |

Fig. 2.5. A table showing whole numbers for use with dividing by 6

Why is the difference of two numbers in the same column a multiple of 6? Why will two numbers in the same column have the same remainder when divided by 6? How is this like working with a calendar? How are the properties of multiplication and addition useful in justifying your answers to these questions? How can we use the quotient, remainder, and division algorithm in making sense of this table, as Essential Understanding 1*f* suggests is possible?

## Percent

Finding a percentage of a number invokes the concept of performing a scalar operation. For example, suppose that we need to find 20 percent of 900. One strategy that we could use would be to partition a quantity representing the 900 into 100 equal-sized pieces and then select 20 of them. In using this strategy, we would be considering *percent* as "so many parts per 100 parts." But in selecting this strategy of partitioning a quantity representing 900 into 100 equal-sized parts and selecting 20, we could also be interpreting 20 percent as the number $20/100$ acting on the quantity representing 900. In this way, we would understand the number $20/100$ as a scalar acting on the quantity that represents 900. We could represent this by the scalar multiplication $0.20 \times 900$. Thus, finding a percent of a number is an instance of scalar multiplication.

An understanding of the role played by a scalar operator is interwoven into an understanding of percents. In working with percents, we must correctly identify the scalar and the quantity on which the scalar acts (see Essential Understanding 1c). A major source of error is incorrectly identifying the appropriate scalar and quantity.

Recognizing the connections between a percentage and a scalar operator can also help students determine when the multiplication of percents is appropriate. If 85% of the population has cell phones and 30% of those cell phone owners have two or more cell phones,

Essential Understanding 1*f*

*Using proper terminology and understanding the division algorithm provide the basis for understanding how numbers such as the quotient and remainder are used in a division situation.*

Essential Understanding 1c

*A situation that can be represented by multiplication has an element that represents the scalar and an element that represents the quantity to which the scalar applies.*

then the product of 0.30 and 0.85, expressed as a percent, is the percentage of the population that has two or more cell phones:

$$0.30 \times 0.85 = 0.255, \text{ or } 25.5\%$$

This calculation has its justification in the observation that if $A$ represents the total number of people, then $0.30(0.85A)$ is the number of people with two or more cell phones. The use of the scalar operation interpretation is critical in giving meaning to this calculation, enabling students to see that 0.30 is a scalar acting on $0.85A$.

Consider how the preceding situation differs from one in which 85% of the population has cell phones and 30% of the population has two or more cell phones. In this new situation, if $A$ represents the total number of people, then $0.85A$ represents the number of people with cell phones, and $0.30A$ represents the number with two or more cell phones. Multiplying the percentages is inappropriate in this case, since both have the same base—that is, the quantity on which the scalar is acting is the same for both.

## Ratio tables

Using ratio tables is a common strategy for solving proportional reasoning problems. For example, suppose that a recipe for spaghetti sauce calls for three thinly sliced onions for 24 ounces of sauce. How much sauce can we make if we have just 2 onions? Or how many onions will we need if we must make 32 ounces of sauce? Each pair of numerical entries in the table in figure 2.6 represents the ratio, or multiplicative relationship, between the two quantities in the recipe.

| Number of onions | 3 | 6 | 12 | 30 | 42 | 2 | ? | | |
|---|---|---|---|---|---|---|---|---|---|
| Number of ounces of spaghetti sauce | 24 | 48 | 96 | 240 | 336 | ? | 32 | | |

Fig. 2.6. A ratio table for the number of onions to number of ounces of spaghetti sauce

Often when solving problems of this nature, we must find other pairs of entries for the ratio table. Such is the case with our problem: how many ounces of sauce should we pair with 2 onions? Or how many onions should we pair with 32 ounces of sauce? Any other pair in the ratio table must reflect the same multiplicative relationship as that between the original pair of entries. Each pair of numbers representing the number of onions and the number of ounces of sauce in the ratio table can be thought of as the result of a scalar multiplication of any other pair of entries in the ratio table. Understanding that applying any scalar factor to a pair in the table produces another pair that belongs in the table is a basis for creating equivalent ratios and proportional reasoning (see Lobato and Ellis [2010]). Determine scalar operators that you need to fill in the cells with question marks in the table in figure 2.6.

*Developing Essential Understanding of Ratios, Proportions, and Proportional Reasoning for Teaching Mathematics in Grades 6–8* (Lobato and Ellis 2010) elaborates on producing equivalent ratios and using proportional reasoning.

## Similarity

Scalar factors play an important role in understanding similar figures in geometry. Similarity can be defined by using concepts of size transformations and isometries. Two polygons A and B are similar if there is a one-to-one correspondence between the vertices of A and B such that the interior angles at corresponding vertices have the same measure and the lengths of sides between corresponding pairs of vertices are proportional by the same scalar factor.

Let's examine what it would mean to have a polygon similar to a quadrilateral with vertices $A_1$, $A_2$, $A_3$, and $A_4$, such as that in figure 2.7a. The vertices of the quadrilateral in figure 2.7b are labeled $B_1$, $B_2$, $B_3$, and $B_4$ in such a way that the angles at $A_1$, $A_2$, $A_3$, and $A_4$ have the same measure as the angles at $B_1$, $B_2$, $B_3$, and $B_4$, respectively, and the length of the corresponding sides between corresponding pairs of vertices, such as the length of the sides between $A_1$ and $A_2$ and between $B_1$ and $B_2$, are proportional by the same scalar factor, 2.

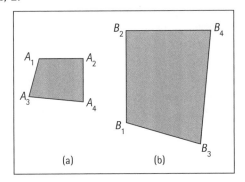

(a)                          (b)

**Fig. 2.7. Similar polygons with a scale factor of 2**

Similar polygons are highly useful in the study of geometry. They have a large number of properties, most of which depend on the scalar relationship between the lengths of the sides of similar figures. In high school, students learn to use matrices to represent geometric figures, and they use an operation called *scalar multiplication* to transform shapes into similar ones through dilations. The preceding discussions of similar figures and ratio tables illustrate how understanding multiplication as a scalar operation is useful in solving problems in many different areas of mathematics (Essential Understanding 1*d*).

Essential ⬅
Understanding 1d

*A scalar definition of multiplication is useful in representing and solving problems beyond whole number multiplication and division.*

# Properties and Algorithms

The definition of multiplication as a scalar operation can serve as an umbrella for many representations for the operation of multiplication of whole numbers. The concept of a scalar operation is also seen in other areas of mathematics when algorithms are involved.

## Operations and their properties

The properties of multiplication and addition play an important role in developing a deep understanding of the algorithms for operations with whole numbers. An understanding of the properties for whole numbers becomes a critical component in developing a deeper understanding of algorithms for operations with rational numbers and in algebra. An understanding of the ideas of algebra depends on both proficiency in and understanding of the procedures needed for computations with rational numbers and manipulations in algebra.

    The use of the associative, commutative, and distributive properties in developing and reasoning about algorithms for multiplication and division of multi-digit whole numbers (Essential Understanding 2a) lays a foundation for understanding how we use those operations with decimals. The properties that justify whole number computational procedures can be used to justify computational procedures with decimals.

    The distributive property of multiplication over addition is fundamental in developing and understanding computational algorithms for whole numbers. This property is also fundamental in understanding the common algorithm for the addition of fractions with the same denominator. For example, suppose that we are finding the sum of $3/7$ and $2/7$. First, we note that $3/7$ is $3 \times (1/7)$, with 3 serving as a scalar applied to $1/7$. Likewise, $2/7$ is $2 \times (1/7)$. We then have

$$\frac{3}{7} + \frac{2}{7} = 3 \times \frac{1}{7} + 2 \times \frac{1}{7} = (3+2) \times \frac{1}{7} = 5 \times \frac{1}{7} = \frac{5}{7},$$

using the distributive property as a key property.

    The multiplication of whole numbers is frequently viewed as unrelated to the multiplication of fractions. When we represent multiplication by the repeated addition, array, or Cartesian product models, it can be difficult for us to connect the multiplication of whole numbers with the multiplication of fractions, such as $2/3$ and $5/7$. Although using an area representation for multiplication might appear to be compatible with the multiplication of both whole numbers and fractions, we always face the issue of why we use multiplication to find area. Further, when we consider applications such as similarity, where scale factors are involved, then repeated addition,

➡️ Essential
Understanding 2*a*

*The commutative and associative properties of multiplication and the distributive property of multiplication over addition ensure flexibility in computations with whole numbers and provide justifications for sequences of computations with them.*

array, area, and Cartesian product models are inadequate to represent the situations involved.

By contrast, a scalar interpretation for the multiplication of fractions provides a way to understand what the product $2/3 \times 5/7$ means. By this interpretation, $2/3$ acts on the quantity described by $5/7$ and changes it. We think of $2/3$ as being $2 \times (1/3)$, so we can interpret the scalar action as first taking $1/3$ of the quantity and then doubling this new amount. Thus, a scalar operation description for the multiplication of fractions provides a meaningful interpretation. Furthermore, an algorithm for the multiplication of fractions follows from the associative and commutative properties of multiplication, as the following example illustrates:

$$\frac{2}{3} \times \frac{5}{7} = 2 \times \frac{1}{3} \times 5 \times \frac{1}{7} = (2 \times 5) \times \left(\frac{1}{3} \times \frac{1}{7}\right) = 10 \times \frac{1}{21} = \frac{10}{21}$$

Note that this representation can also explain the common procedure for multiplying fractions, if we simply reorder the steps:

$$\frac{2}{3} \times \frac{5}{7} = 2 \times \frac{1}{3} \times 5 \times \frac{1}{7} = (2 \times 5) \times \left(\frac{1}{3} \times \frac{1}{7}\right) = \left(\frac{2 \times 5}{3 \times 7}\right) = \frac{10}{21}$$

We determine the product $1/3 \times 1/7$ by using the scalar description of taking one-third of one-seventh of a quantity. Barnett-Clarke and colleagues (2010) discuss various representations for making sense of the multiplication of fractions such as $2/3$ and $5/7$. Because of the relationships between fractions and decimal numbers, a scalar description of multiplication also provides a meaningful interpretation of the multiplication of decimal numbers.

Because teachers usually introduce multiplication as repeated addition and frequently reinforce this description, students often develop the belief that "multiplication makes bigger." Along with this belief, they commonly develop a corresponding belief about division—that "division makes smaller." Although these generalizations are valid for whole numbers, they do not apply to operations with real numbers. Graeber (1993) elaborates on how students develop these limited conceptions. Lannin and colleagues (forthcoming) explain the mathematical reasoning that underlies the use of generalizations like these. Because a scalar operation affects the size of a quantity, interpreting multiplication—and division through its inverse relationship with multiplication—in this way challenges these two beliefs. We can come up with examples that demonstrate how multiplication and division can change the magnitude of a quantity by increasing or decreasing its size. Consider the following questions, for instance:

- What number can you multiply 356 by to get a product between 50 and 100?

For a full discussion of operations on fractions and decimals as an extension of operations on whole numbers, see *Developing Essential Understanding of Rational Numbers for Teaching Mathematics in Grades 3–5* (Barnett-Clarke et al. 2010).

For a discussion of generalizations, see *Developing Essential Understanding of Mathematical Reasoning for Teaching Mathematics in Prekindergarten–Grade 8* (Lannin et al., forthcoming).

• What number can you divide 365 by to get a quotient greater than 1000?

Answering questions such as these can help develop number sense and estimation skills.

## Exponents

The properties of multiplication are also keys to developing an understanding of natural number exponents. The generalized associative property of multiplication serves to justify many of the properties of exponents. For example, we can justify $a^3 \times a^4 = (a \times a \times a) \times (a \times a \times a \times a) = a \times a \times a \times a \times a \times a \times a = a^7$. We can provide a similar justification for $(a^3)^4 = (a \times a \times a) \times (a \times a \times a) \times (a \times a \times a) \times (a \times a \times a) = a^{12}$. We need both the generalized commutative and associative properties to justify the use of exponents with indicated products such as the following:

$$(a \times b)^3 = (a \times b) \times (a \times b) \times (a \times b)$$
$$= (a \times a \times a) \times (b \times b \times b) = a^3 \times b^3$$

Also, the powers of a natural number, such as the powers of 10, serve as an example of *exponential growth*. In contrast to an *arithmetic sequence*, formed by adding a fixed amount to the previous term to get the next term, a *geometric sequence* is formed by multiplying each term by the same fixed amount to get the next term. One example of such a sequence is 1, 2, 4, 8, 16, 32, ..., where each term in the sequence is double the previous term. This sequence models many growth patterns in nature. Another example of a geometric sequence is a sequence where each term is ten times the preceding term. Starting with 4, for instance, the next terms are 40, 400, 4,000, 40,000, .... A graph of such a sequence on the Cartesian plane as the ordered pairs (1, 4), (2, 40), (3, 400), and so forth, does not match the linear pattern of an arithmetic sequence but instead matches a curve which grows more and more rapidly; this is an example of an exponential function. Both types of sequences, arithmetic and geometric, are important ideas in the study of combinatorics and calculus.

## Conclusion

Multiplication and division ideas are introduced in the primary grades. Students begin to develop an informal understanding of multiplication and division through many activities such as skip counting and sharing. Even though a focus of learning about multiplication in grades 3–5 is on developing computational proficiency, students also begin to see the usefulness of multiplication as a problem-solving instrument in many other situations. As students learn to recognize various situations as multiplicative, they begin to form an idea of multiplication as a powerful mechanism for representation and reasoning.

This solidly rooted and growing understanding of multiplication and division in grades 3–5 lays the foundation for further understanding of these operations in other number domains, such as fractions and decimals. Furthermore, this understanding is an essential component in developing an understanding of algebra and proportional reasoning, including that used in geometry.

# Challenges: Learning, Teaching, and Assessing

Teachers need to develop understanding of the big ideas and essential understandings related to multiplication and division. Otherwise, they often use an instructional approach that does not promote a flexible, growing understanding in their students. Consider the following discussion between two teachers.

Ms. Avery and Ms. Wojak are discussing their beliefs about what it means for the students in each of their respective classrooms to have an understanding of the concepts and number facts related to multiplication and division. Ms. Avery believes that if students memorize their basic facts, they will be able to solve problems. She believes that a teacher's role is to help students get the correct answer and that getting the correct answer shows understanding. She explains her approach to Ms. Wojak in more detail:

> My students have been studying fact families so that they can see the relationship among the three numbers in multiplication and division facts. If I show them a triangular card with the numbers 3, 4, and 12 on it, with one number in each corner, they know that $3 \times 4$ is 12, that $4 \times 3$ is 12, that $12 \div 4$ is 3, and that $12 \div 3$ is 4. We play a game using fact-family cards in an "Around the World" format, and we talk about how the numbers on the card are related, just like the people in a family. The students come to understand multiplication and division through this game.

Ms. Wojak thinks that an approach that emphasizes memorization does not necessarily build in-depth understanding. She elaborates on her own very different approach:

> That's interesting. For my students to understand multiplication and division, I select a number triple such as 5, 6, and 30, and I create a word problem with those numbers, using a context familiar to my students so that they can make sense of the

→ Essential
Understanding 1c
A situation that can
be represented by
multiplications has
an element that
represents the scalar
and an element
that represents the
quantity to which the
scalar applies.

numbers. For example, I might begin by saying, "If there are 5 children and each gets 6 balloons, how many balloons are there?" I want to determine if they think of five 6s and count by sixes or if they realize they can use the commutative property and think six 5s and then count by fives to get an answer. Or some may use the distributive property and known facts to reason, saying something like, "If I know three 5s are 15, then six 5s are 15 and 15, which is 30." Later in the lesson, I might ask, "If there are 30 cookies and 6 children, how many cookies can each child get if they each get the same number?" When students realize that division and multiplication have an inverse relationship and know that $5 \times 6$ is 30, they can find that 30 cookies shared fairly among 6 children is 5 cookies each. Then I am confident that they understand multiplication and division.

→ Essential
Understanding 2a
The commutative
and associative
properties of
multiplication and
the distributive
property of
multiplication over
addition ensure
flexibility in
computations with
whole numbers and
provide justifications
for sequences of
computations with
them.

Note that this approach has the potential to help Ms. Wojak's students develop robust understanding—while allowing Ms. Wojak to assess their understanding in relation to several essential understandings. After giving her students the problem involving the 5 children with 6 balloons apiece, Ms. Wojak can observe them as they think about the elements in this multiplicative situation and identify one as the scalar and the other as the quantity on which the scalar acts (Essential Understanding 1c). She can note whether or not the students make use of the distributive property to compute flexibly (Essential Understanding 2a). She can then reverse the situation involving the number triple 5, 6, 30, presenting the students with the new problem of sharing 30 cookies fairly among 6 children, to see if they understand the inverse relationship between division and multiplication (Essential Understanding 1e).

→ Essential
Understanding 1e
Division is defined
by its inverse
relationship with
multiplication.

Ms. Wojak continues thinking about how she plans instruction for her students. She knows that whenever she is introducing a concept, she usually starts with a word problem to determine what mathematics the students know. She listens for different reasoning strategies. For the cookie problem, she thinks that one student might say, "I started with 30 cookies and gave each person 1. Then 6 were gone from the 30, and I had 24." The student might continue double counting: "2 groups of 6, 18 left; 3 groups of 6, 12 left; 4 groups of 6, 6 left; 5 groups of 6, 0 left. So each person gets 5 cookies." If at that point no one knew how to represent the situation symbolically, Ms. Wojak thinks she would write the equation $5 \times 6 = 30$ on the board, reading the symbol "$\times$" as "groups of" if those were the words that the student used. This equation would mean that 5 acts as the scalar, or operates on the quantity 6 (Essential Understanding 1a). Ms. Wojak explains that she would encourage a variety of different solution strategies during the lesson (Essential Understanding 1b).

→ Essential
Understanding 1a

In the multiplicative
expression $A \times B$,
A can be defined as a
scaling factor.

Ms. Wojak believes that to develop her students' understanding of ideas and facts related to multiplication and division, she must recognize their reasoning and help them to represent it symbolically, exactly following the thinking processes that they have used. She makes opportunities for students to share ideas during strategy sessions that elicit several different reasoning processes that represent most of the ways in which students in the class are thinking.

Ms. Wojak meets individually with students who are not yet using a strategy to help them learn to model a situation directly with pictures or manipulative materials of their choice. Moreover, Ms. Wojak knows that repeated subtraction is just a beginning division strategy for most students and that to make sense of larger numbers, they will also need to make sense of more sophisticated strategies, such as those that use the right distributive property (Essential Understanding 2*b*). Her goal is for students to develop a rich and growing understanding of the concepts and strategies related to multiplication and division and to develop an efficient algorithm based on those understandings.

Both Ms. Avery and Ms. Wojak have an idea of what *understanding* means. Each teacher believes strongly that her students must understand concepts and facts related to multiplication and division. Yet, students in the two classes will develop different mathematical understandings.

Ms. Avery's aim is for her students to develop symbolic proficiency, while Ms. Wojak is striving to help her students develop a more robust understanding of multiplication and division. To accomplish her goal, Ms. Wojak must have a deep understanding of the essential understandings so that she can respond to the different reasoning that students use to represent and solve problems and help them develop more efficient solution strategies.

## Teaching for Robust Understanding

The scenario with Ms. Avery and Ms. Wojak brings up the issue of what each teacher believes is the meaning of *understanding*. For some teachers, an indicator of understanding might be proficiency with symbolic procedures. Such teachers might ask their students questions like, "What do you do next?" and, "Where do you write that number?"

As Ms. Avery and Ms. Wojak continue their discussion about understanding of multiplication and division, they begin to identify more specific indicators of understanding. When Ms. Avery asks a student to multiply two two-digit numbers, such as 24 and 35, for example, she expects the student to say (and do) something like, "OK, 4 times 5 is 20, so I put down the zero and carry the 2; 4 times 3 is 12 and 2 more is 14, so I write down 140, and then...." This

**Essential Understanding 1*b***

*Each multiplicative expression developed in the context of a problem situation has an accompanying explanation, and different representations and ways of reasoning about a situation can lead to different expressions or equations.*

**Essential Understanding 2*b***

*The right distributive property of division over addition allows computing flexibly and justifying computations with whole numbers, but there is no left distributive property of division over addition and no commutative or associative property of division of whole numbers.*

"explanation" does not provide evidence that a student has a rich understanding of the mathematical concepts involved in two-digit multiplication. It merely describes a symbolic procedure that might reflect a limited understanding.

Teachers like Ms. Wojak, by contrast, focus on the thinking that students are using, and they ask questions to help the students enhance their understanding as they develop other ways to represent and solve problems. These teachers' questions are designed to push behind procedures and probe students' understanding of their steps:

- Why do you put down a zero here?

- What does the 2 that you carried here represent?

- Why do you add 10 and 2, when this is a multiplication problem?

**Big Idea 2**

*The properties of multiplication and addition provide the mathematical foundation for understanding computational procedures for multiplication and division, including mental computation and estimation strategies, invented algorithms, and standard algorithms.*

These types of questions address the "whys" of the process. They draw on the teacher's understanding of Big Idea 2 rather than his or her familiarity and efficiency with procedures. Furthermore, instead of eliciting from students a list of memorized procedural steps, these questions compel students to reason about, and build their understanding of, concepts. Teachers who focus on this kind of understanding of multi-digit multiplication help their students make connections among a variety of mathematical concepts. Let's examine some of the ways in which a teacher with deep understanding might help students learn to calculate a product like 24 × 35 with a flexible, robust, and growing understanding.

## Beginning with a word problem

A good way to begin is to create a word problem to assess strategies that students might intuitively use. For example, Mr. Robinson creates the following word problem and shares it with his students to gain insight into their understanding of the multiplication of two two-digit numbers:

> Each of 24 students in the fourth grade brought 75 cents for a school party. How much money did the fourth grade then have for the party?

Mr. Robinson observes that one student, Olivia, writes the following in her first attempt to solve the problem:

$$2 \times 75 = 150$$
$$4 \times 75 = 150 + 150$$
$$8 \times 75 = 300 + 300$$
$$16 \times 75 = 600 + 600 = 1200$$

Mr. Robinson sees that Olivia knows that she needs only 8 more groups of 75. Because 16 × 75 = 1200 and 8 × 75 = 600,

24 × 75 = 1200 + 600 = 1800 cents, or $18.00. Questioning leads Mr. Robinson to believe that Olivia's understanding does not go much beyond enabling her to use a shortcut for repeated addition, but he considers this approach a good beginning. He wants to make sure that Olivia recognizes connections between what she is doing and the distributive property (Essential Understanding 2a), seeing, for example, that 24 × 75 is the same as (16 + 8) × 75, or (16 × 75) + (8 × 75).

## Using carefully selected computations

A second way of building students' understanding is by questioning them about carefully chosen computations. After creating the problem of the fourth graders who brought 75 cents apiece for a party, Mr. Robinson might develop a series of computations and accompanying questions for his students to reason about to help them build their understanding. Suppose that Mr. Robinson selects the following sequence of computational questions:

- How are 2 × 7 and 2 × 70 related?

  Mr. Robinson knows that students need to understand that the second product is not formed by "adding a zero" to the first product, but by becoming "10 times greater" than the first product. The product changes by a factor of 10, from 14 to 14 tens, or 140. Mr. Robinson knows that students can intuitively use the associative property to show how the factor of 10 emerges, and he introduces the formal name of this property after students use it in their reasoning:

  $$2 \times 70 = 2 \times (7 \times 10) = (2 \times 7) \times 10 = 14 \times 10$$

- How are 2 × 70 and 2 × 75 related?

  Although they do not yet know the distributive property, students use it informally as they draw on their understanding of multiplication and number decomposition to show the connection:

  $$2 \times 75 = 2 \times (70 + 5) = (2 \times 70) + (2 \times 5)$$

- How are 2 × 70 and 20 × 70 related?

  Because 20 is ten times 2, 20 × 70 is ten times 2 × 70:

  $$20 \times 70 = (2 \times 10) \times 70 = (10 \times 2) \times 70 = 10 \times (2 \times 70)$$

- How are 2 × 75 and 20 × 75 related?

  $$20 \times 75 = (2 \times 10) \times 75 = (10 \times 2) \times 75 = 10 \times (2 \times 75)$$

- How are 20 × 75 and 24 × 75 related?

  $$24 \times 75 = (20 + 4) \times 75 = (20 \times 75) + (4 \times 75)$$

Essential ⬅
Understanding 2a

*The commutative and associative properties of multiplication and the distributive property of multiplication over addition ensure flexibility in computations with whole numbers and provide justifications for sequences of computations with them.*

Mr. Robinson makes use of properties and computational fluency to generate questions to help his students to develop their flexibility in computation (Essential Understanding 2a), while simultaneously emphasizing correct mathematical terminology that contributes to mathematical understanding (Essential Understanding 1f).

## Working with an area model

A third way of building students' understanding applies their knowledge of area as length times width. If the length of a rectangle is 75 units and the width is 24 units, students can develop a rectangular model to represent the area, such as that shown in figure 3.1. Such an area model provides a geometric representation of the distributive property.

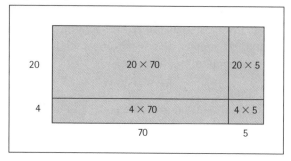

Fig. 3.1. Using an area model to illustrate a procedure for multiplying two-digit numbers

A student who represents her thinking with an area model, which might develop naturally from her use of base-ten materials to explain her thinking, could say something like the following:

> First, I can think about 20 + 4 times 70. I know that 20 × 70 is 1400 because 20 × 70 is ten times 2 × 70. And 4 times 70 is 280 because 4 × 70 is 10 times 4 × 7. Then 20 × 5 is 100, and 4 × 5 is 20. By doing this in parts and using what I know, I can justify that 1400 + 280 + 100 + 20 gets the correct answer of 1800.

When students recognize and can use connections among parts of the area model and the symbolic procedures for multiplying, they demonstrate a rich mathematical understanding.

## Looking—and Listening—for Understanding

As students extend their knowledge, the teacher's challenge is to provide opportunities to develop more in-depth understandings. Consider Ms. Wojak's approach, as she explains it to Ms. Avery.

Ms. Wojak relates to Ms. Avery an earlier conversation in her classroom. One student, Rena, described her strategy for solving a problem that required sharing 30 cookies among three people. Rena said, "OK, 3 × 10. You give them 10 because you have 30 cookies. If you take 10, 10, and 10, that equals 30. So, there are three people; each can get 10. I switched it around, so it was 30 and 3, and that equaled 10." Ms. Wojak explains that when she asked Rena what she did with the 30, Rena said, "I split 30 into tens. I knew 3 × 10 is 30, so 30 has three tens." Ms. Wojak tells Ms. Avery that at that point, she asked Rena, "What operation symbol would represent that idea of 'splitting into equal parts,' and how would we represent your thinking with an equation?" Rena wasn't sure, so Ms. Wojak posed the following problem:

> If 18 peanut butter cookies are shared among three people, how many cookies will each person get?

Rena said, "Six. I know that 3 × 6 is 18, and if I divide 18 into 3 groups, I will have 6." This time, Rena used the word *divide* instead of *split*, and Ms. Wojak represents her thinking by writing 3 × 6 = 18 and 18 ÷ 3 = 6.

The classroom conversation continued with other students using their understanding of multiplication to derive division sentences. Ms. Wojak explains to Ms. Avery that her students were beginning to develop an understanding of the inverse relationship between multiplication and division (Essential Understanding 1*e*), and she emphasized the terms *multiplication*, *division*, *factor*, *multiple*, and *product* throughout the lesson (Essential Understanding 1*f*). At the end of the discussion, Dashawn used the right distributive property over division to explain his reasoning (Essential Understanding 2*b*), explaining, "Twenty-one divided by 3 is the same as 15 divided by 3 plus 6 divided by 3." Ms. Wojak asked him to write the equation that he had verbalized:

$$21 ÷ 3 = (15 ÷ 3) + (6 ÷ 3)$$

# Hallmarks of Teaching with Understanding

Classroom situations inevitably bring attention to a variety of issues in teaching about multiplication and division, including the following:

- The use of "tricks" in computation
- Opportunities for rich mathematical tasks
- The need for mathematical terms and definitions

Essential
Understanding 1*e*

*Division is defined by its inverse relationship with multiplication.*

Essential
Understanding 1*f*

*Using proper terminology and understanding the division algorithm provide the basis for understanding how numbers such as the* quotient *and* remainder *are used in a division situation.*

Essential
Understanding 2*b*

*The right distributive property of division over addition allows computing flexibly and justifying computations with whole numbers, but there is no left distributive property of division over addition and no commutative or associative property of division of whole numbers.*

When teachers see their students' experiences of multiplication and division through their own understanding of the big ideas and essential understandings, they discover ways to turn these three issues into new opportunities for their students to develop their understanding of multiplication and division.

## Explaining novel procedures

Students who focus on getting correct answers quickly, without reasoning, come to believe that mathematics requires mostly "tricks" and memorization of procedures. This kind of learning sacrifices understanding. A learning environment that focuses on understanding presents students with no tricks, shortcuts, or mysterious procedures, and the teacher's challenge is to create opportunities for students to use and develop understanding by exploring why "tricks" or memorized procedures make sense.

Consider the following procedure for multiplying by 9. If you number your fingers left to right from 1 through 10 as shown in figure 3.2, and hold finger 2 down, you have one finger to its left and eight fingers to its right. This arrangement "shows" the product of 9 and 2 as 18. If you hold finger 6 down, you have 5 fingers to its left and 4 fingers to its right. This arrangement "shows" that $9 \times 6$ is 54.

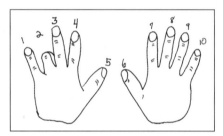

Fig. 3.2. Representing $9 \times 2$ by using a common shortcut that need not be a meaningless procedure for multiplying by 9

What understanding can using this procedure reinforce? What mathematics can we bring to this procedure to make sense of it? One way to reason about and justify the procedure follows:

$9 \times n = (10 - 1) \times n = 10n - 1n$      By using what we know and the distributive property

$9 \times n = 10n - 1n = 10n - 10 + 10 - 1n$      By adding and subtracting 10

$9 \times n = 10n - 10 + 10 - 1n = 10(n - 1) + (10 - n)$      By using the distributive property again

In the hand procedure for multiplying $9 \times n-$

- $n$ is the number of the finger held down;

- $n - 1$ represents the number of fingers to the left of this finger;

- $10 - n$ represents the number of fingers to its right; and

- $10(n - 1) + (10 - n)$ is equal to $9 \times n$.

Thus, our use of the distributive property provides a mathematical explanation for this "trick" with fingers (Essential Understanding 2*a*).

Teachers who teach for conceptual understanding might explore with their students why tricks, shortcuts, and seemingly mysterious procedures work. It is through such exploration that students delve into mathematics and remember it in the long run. It is important to note that the reasoning processes behind the tricks, shortcuts, and procedures may take time to develop. However, these reasoning processes can reinforce students' understanding of mathematics encountered outside the classroom and within previous mathematics lessons. In fact, students can have quite a bit of fun determining how they can use the mathematics that they understand to explain "tricks" to people who won't take time to think!

## Using worthwhile tasks

Teachers can create tasks to motivate students to reason, promoting their understanding in many ways. For example, a fifth-grade teacher, Mr. Charles, puts up an interactive bulletin board covering the entire back wall of his classroom. The "four 4s" board displays the numerals representing the numbers from 1 to 100. The students' objective is to find as many expressions as they can for each number. They can use any operations but only four 4s. Some examples for 1 and 60 follow:

$$\frac{4 \times 4}{4 \times 4} \qquad \frac{44}{44} \qquad 4 \times 4 \times 4 - 4 \qquad 44 + 4 \times 4$$

Several expressions are possible for each number. Mr. Charles's students discover that order of operations, number relationships involving multiplication and division, and the properties of operations all play roles in finding alternative ways of expressing the same number. Periodically, Mr. Charles discusses these ideas to help his students understand the properties and relationships.

Taking advantage of a different setting, Carey (1992) suggests using children's literature to motivate students to solve problems. Using the book *The Patchwork Quilt* (Flournoy 1985), Carey proposes tasks involving multiplication and division concepts. By carefully wording problems, she encourages students to use multiple representations for multiplication and division situations. For example, an array representation can model one way of thinking about the problem, "How many squares will Tanya need if she makes a quilt that has 6 rows with 4 squares in each row?" (Carey 1992, p. 200).

Essential
Understanding 2*a*

*The commutative and associative properties of multiplication and the distributive property of multiplication over addition ensure flexibility in computations with whole numbers and provide justifications for sequences of computations with them.*

By thoughtfully selecting the numbers for each problem, Carey encourages thinking about factors, commutativity, and concepts involving the relationship of area and perimeter. An extension task on scale drawing makes connections to the concept of scalar multi-plication. If the length and width of a rectangle are both doubled, as shown in figure 3.3, what is the effect on the perimeter of the shape and the area of the rectangular space? Thiessen (1998, 2004) details additional connections to literature that incorporate concepts related to multiplication and division.

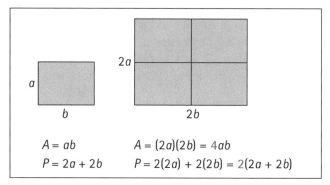

Fig. 3.3. Diagrams and expressions illustrating the effects of doubling the length and width of a rectangle on its area and perimeter

Topics from units that students are working on in science, so-cial studies, and art can also provide contexts for reasoning about multiplication and division, as can topics related to students' school life in general. For example, consider the following situation:

> If 35 third-grade students bring $25 each to purchase a school jacket, how much money is that? If 45 fourth-grade students bring $25 each for the same purpose, how many more dollars do the fourth-graders bring for jackets?

Students should be able to answer the second question with mini-mal calculations and no paper and pencil.

Also consider presenting problems to students that connect to their interests outside school. The following is just one example:

> For a Super Bowl, _____ rings were purchased by the winning team. If each ring costs $ _____ , how much was the bill for all of the rings?

Students could research the situation for homework and bring in actual data so that the problem to be solved reflects the current real-life situation. Strategy sharing can also be a mental mathematics activity. In Mr. Lee's class, the students found that 150 rings were purchased at a cost of $5,000 each. One student, David, reasoned that 100 rings would cost 100 × 5,000, or $500,000. Fifty more rings would cost half as much, or $250,000. So David came up with a total cost of $750,000.

Information that students might use for the Super Bowl problem can be found at http://www .nola.com/superbowl /index.ssf/2010/06 /new_orleans_saints_ super_bowl_11.html.

Meanwhile, another student, Marianne, began by considering that 5 × 15 is 75. She knew that the number of dollars needed would be ten thousand times greater than 75. So she reasoned that the total cost would be $750,000. Her calculation can be justified by using properties of whole number multiplication and addition:

$$10,000 \times (5 \times 15) = 1,000 \times (5 \times 15) \times 10 = (1,000 \times 5) \times (15 \times 10) = 5,000 \times 150$$

When students share reasoning strategies on a daily basis, they have opportunities to build on their own strategies. For all students, an important part of the sharing of strategies is questioning and explaining as they develop a deeper and richer understanding.

## Using appropriate language and definitions

As students progress from grade 3 to 4 to 5 and continue to discuss strategies, teachers should use the same terms to ensure effective communication about the mathematics that the students are learning. Confusion among students can result if one teacher says *turn-arounds* and another says *commutative*. If a teacher does not use a mathematical term precisely, students may use the concept incorrectly. For example, they may apply the commutative property of multiplication to the operation of division—an error that teachers often overhear students making in discussions about division.

Teachers can help students develop communication and reasoning skills by referring to terms and definitions that the students are familiar with but have neglected to use as part of their shared reasoning process. Consider the following scenario. Kobe is confused, so his teacher, Mr. Dean, engages him in a discussion:

*Kobe:* I can't remember if 5 ÷ 0 = 0, or 0 ÷ 5 = 0. Or are they both 0?

*Mr. Dean:* What do we mean when we say that we *divide*?

*Kobe:* [*Thinking*] Let's see... If 12 ÷ 3 = 4, then 12 = 3 × 4... So, for 5 ÷ 0 = 0, then 5 = 0 × 0. That doesn't make sense. Let's try 0 ÷ 5 = 0, so 0 = 5 × 0. That makes sense. So 0 ÷ 5 = 0 makes sense, but 5 ÷ 0 = 0 does not.

*Mr. Dean:* [*Summarizing*] We call 5 ÷ 0 *undefined* because division by zero does not make sense when we use the definition of division.

In this scenario, the teacher helped the student by using an idea that the class had studied to help make sense of division involving zero. The teacher also introduced the term *undefined* so that the student could extend his mathematical vocabulary. Mathematical reasoning and discourse are enhanced when teachers use correct terms and help students use definitions that their previous work has helped them to understand.

# Making Time to Foster Understanding

Developing understanding requires not only worthwhile tasks and appropriate language but also time. Consider the time that Mr. Fuentes devotes to helping his students develop an understanding of division computation. He presents the following problem to his class:

> Mr. Fuentes has 698 pencils. He is going to give each of the fifth graders a box of 12 pencils. How many boxes will Mr. Fuentes be able to fill?

Mr. Fuentes makes sure all of the students understand the problem and encourages them to get any materials that they need to work on it. Then he walks around the room observing students' work. When Mr. Fuentes approaches Alyssa, he sees that she has written the expressions that appear in figure 3.4.

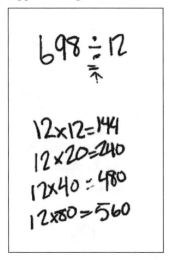

Fig. 3.4. Alyssa's initial work on Mr. Fuentes's problem

Essential Understanding 1e

*Division is defined by its inverse relationship with multiplication.*

Her written work makes Mr. Fuentes think that Alyssa understands that division has an inverse relationship with multiplication (Essential Understanding 1e) and recognizes that multiplying by tens is efficient. He notices her calculation error, but without calling attention to it, he says, "I see what you've done. Explain your reasoning to me. What number are you trying to get to?" Alyssa explains that she is trying to get to 698 and that she needs to figure out how many multiples of 12 are in that number. She knows that $10 \times 12$ is 120, and she uses that product to derive $12 \times 20$ and doubles her new product to get $12 \times 40$. She says, "And that's still not high enough, so I'm going to do $12 \times 80$." She miscalculates again and says, "That's still not high enough." Mr. Fuentes says,

"Wait a minute, if 12 × 40 is 480, then how did you get 12 × 80 is 560?" Alyssa thinks and says, "That's wrong, I forgot to double the 4... That would be too much."

At this point, Mr. Fuentes wants to use Alyssa's thinking, in addition to her knowledge of both basic facts and multiplying by 10s and 100s, to help her to develop a more efficient algorithm. He begins a dialogue with Alyssa to help her make connections from the mathematics that she knows to the mathematics he wants her to know. He uses a series of questions as he writes out significant numbers from her thinking process:

*Mr. Fuentes:*  You want to know how many 12s are in 698 and you want to get there quickly.

*Alyssa:*  [*Nods.*]

*Mr. Fuentes:*  And you know that for every ten 12s, it's 120, so let's work with this. How many 12s are in 698, just looking at it? How many times are you going to get 120 out of there?

*Alyssa:*  [*Estimating*] Four.

*Mr. Fuentes:*  But you already did 40 times 12 and didn't get to 698.

*Alyssa:*  [*Estimating*] Maybe 50.

*Mr. Fuentes:*  So, if you do 50 × 12, what will that get you?

*Alyssa:*  [*Hesitating.*]

*Mr. Fuentes:*  [*Waits before asking another question*] What do you know about 5 × 12?

*Alyssa:*  It's 60.

*Mr. Fuentes:*  So does that help with 50 × 12?

*Alyssa:*  I think so. It's times 10 more, so that would be 600!

*Mr. Fuentes:*  I love that "times 10"—you're not just going to add a zero! So now, look, you're almost there. What do you know at this point?

*Alyssa:*  I only have 98 more to use.

*Mr. Fuentes:*  Yes, so what's a good estimate?

*Alyssa:*  Five 12s are 60. So I would have 38 more to get to 98.

*Mr. Fuentes:*  So, what do you know about 38?

*Alyssa:*  Well, 3 × 12 is 36.

*Mr. Fuentes:*  So, what do you know about this problem for an answer to how many boxes Mr. Fuentes can fill?

*Alyssa:*          It's 58 boxes.

*Mr. Fuentes:* Why?

*Alyssa:*          Because 50 × 12 is 600, then I did 5 × 12 is 60,
                   and then I did 3 × 12, which is 36.

*Mr. Fuentes:* Will all the pencils be in a box?

*Alyssa:*          No, there would be 2 left.

*Mr. Fuentes:* So, if Mr. Fuentes has to have all the pencils in a
                   box, how many boxes will he need?

*Alyssa:*          Um, 59.

*Mr. Fuentes:* Excellent.

At this point, Mr. Fuentes gives Alyssa another problem to assess whether she is beginning to develop a more efficient strategy:

> If we have 8,492 pencils and we put 12 in a box, how many
> boxes exactly will we need?

Alyssa begins writing; her work appears in figure 3.5.

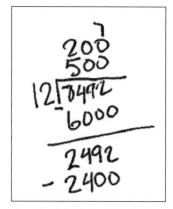

Fig. 3.5. Calculations involved in Alyssa's solution to the second problem that Mr. Fuentes poses

Mr. Fuentes can see that after Alyssa has written 8,492, she
begins to subtract groups of 12. She seems to remember that 5 ×
12 is 60 and then reasons that 500 × 12 is 60 hundreds, or 6,000.
Alyssa subtracts 6,000 from 8,492 to get 2,492, writing 500 above
to keep track of the number of groups of 12 in her quotient. She
realizes that she is left with 2,492 and stops. Mr. Fuentes begins his
questioning:

*Mr. Fuentes:* [*Pointing to the divisor of 12 and the difference of
                   2492*] Twelve and 2,492—I see a number that should
                   pop right into your head; what is it?

*Alyssa:*          Twenty-four popped into my head because it is the
                   double of 12, so 20 pops into my head, 20 × 12 is

240, and 200 × 12 is 2,400. I have 92 left. OK, and I know that 10 × 12 is 120, and that's too much. And 9 × 12 is 108. But I can take away 12 to get 96, and then take away 12 to get 84, so 7 × 12 is 84.

*Mr. Fuentes:* How are you going to use that idea?

*Alyssa:* I'll take that away from the 92 [*writes 84 under 92 on the board and continues writing*], and 92 – 84 = 8, so that would be 8 left.

*Mr. Fuentes:* So what do you conclude as the answer to the number of boxes that we will need?

*Alyssa:* [*Adding the partial quotients and apparently putting the remaining 8 pencils in a separate box*] It's 708!

Mr. Fuentes understands that knowledge of the multiplication facts connected to 12 is helpful to his students' development of this efficient strategy. Furthermore, Mr. Fuentes realizes that his students are developing a fuller understanding of division as they think about a more efficient strategy. He also knows that helping students acquire this rich understanding takes time.

# Embedding Assessment in Instruction

To assess a student's understanding of multiplication and division, a classroom teacher needs to consider each observation, discussion, end-of-unit assessment, work sample, and problem-solving session thoughtfully and critically. Teaching and assessing for understanding are inextricably related. Teachers need to probe students' understanding consistently as they observe them and engage them in dialogue. Through observation, effective teachers can gain critical information about their students' understanding.

Consider how a third-grade teacher, Mr. Henry, uses questioning as a means to assess a student's understanding of division. Mr. Henry gives his students the following problem:

If 3 friends share 18 cookies, how many cookies does each get?

Mr. Henry notes that one student, Tierek, who has begun representing the situation with cubes, uses familiar facts and relationships to get close to the answer. Then he adjusts to get the exact answer.

Because Mr. Henry is interested in using his students' level of understanding as a means of informing instruction, he questions Tierek, probing his reasoning until he is confident that he has identified the depth of Tierek's understanding. Notice how Mr. Henry bases his questions on what Tierek does and says in solving the problem:

*Mr. Henry:* I noticed that first you grouped the cubes in 5s. Why

**Big Idea 1**

*Multiplication is one of two fundamental operations, along with addition, which can be defined so that it is an appropriate choice for representing and solving problems in many different situations.*

➡ **Essential Understanding 1c**
*A situation that can be represented by multiplication has an element that represents the scalar and an element that represents the quantity to which the scalar applies.*

➡ **Essential Understanding 1e**
*Division is defined by its inverse relationship with multiplication.*

➡ **Essential Understanding 2a**
*The commutative and associative properties of multiplication and the distributive property of multiplication over addition ensure flexibility in computations with whole numbers and provide justifications for sequences of computations with them.*

|           |                                                                                 |
|-----------|---------------------------------------------------------------------------------|
|           | did you do that and then stop after you'd built three groups?                    |
| *Tierek:*    | Because I know how to count by 5s, I thought I'd grab them in 5s.             |
| *Mr. Henry:* | You also know that three groups of 5 is 15, so why did you check how many cubes you had out for the cookies by counting 5, 10, 15? |
| *Tierek:*    | Well, I guess I just had to double check.                                     |
| *Mr. Henry:* | I see. After that, you took out only three more unit cubes and added one to each pile. Why did you stop after that? |
| *Tierek:*    | Because I knew that I had only 15 cookies there going to 3 kids—the piles are the kids—and there are supposed to be 18 cookies and 3 kids. So if I added 3 more it would make the eighteen. The answer is 6. |

Mr. Henry poses his questions to gain understanding of Tierek's thinking. Knowledge of a student's thinking, together with a deep understanding of multiplication and division, enables a teacher to create learning tasks that provide opportunities for the student to make sense of representations and reasoning strategies.

Jotting down key instances of students' thinking, their use of mathematical vocabulary, and their work with models or symbolic procedures for representing multiplication and division provides an invaluable written account of how a student's understanding develops over time. Perhaps Tierek has never before used groups of 5—something he knows well—to get close to the answer so he can adjust his work to get the exact answer. His thinking is an example of how students make informal use of the distributive property as they begin to understand the mathematics more formally.

As Mr. Henry listens and draws on his understanding of the big ideas and essential understandings of multiplication and division, he can ask a variety of questions—each based on an essential understanding that he has developed—to assess his students' understanding, including the following:

- Do they understand when to multiply or divide? (Big Idea 1)

- Do they understand how to represent various situations involving multiplication and division? (Essential Understanding 1c)

- Do they understand that multiplication can be used to help them solve division problems? (Essential Understanding 1e)

- Do they understand the properties of multiplication and know when to use them efficiently? (Essential Understanding 2a)

The answers to these questions can help Mr. Henry decide if his students need additional experience solving routine word problems with a variety of structures. These problems can help his students gain a better understanding of when to divide and give them opportunities to use multiplication to help solve division problems.

# Conclusion

The task of teaching multiplication and division concepts is a challenging one. Because these concepts provide a foundation for mathematics far beyond grades 3–5, it is critical that students have opportunities to develop the necessary understanding for future mathematics learning.

This chapter presents several aspects of effective instruction that can help students develop such a foundation. It is important that teachers understand both a variety of strategies as well as how students' thinking can be developed to levels that reflect a flexible and growing understanding. Teachers' knowledge of mathematics is critical to this process because their deep understanding can serve as a conduit through which students develop their own robust understanding.

# References

Barnett-Clarke, Carne, William Fisher, Rick Marks, and Sharon Ross. *Developing Essential Understanding of Rational Numbers for Teaching Mathematics in Grades 3–5.* Essential Understanding Series. Reston, Va.: National Council of Teachers of Mathematics, 2010.

Bates, Tom, and Leo Rousseau. "Will the Real Division Algorithm Please Stand Up?" *Arithmetic Teacher* 33 (March 1986): 42–46.

Blanton, Maria, Linda Levi, Terry Wayne Crites, and Barbara J. Dougherty. *Developing Essential Understanding of Algebraic Thinking for Teaching Mathematics in Grades 3–5.* Essential Understanding Series. Reston, Va.: National Council of Teachers of Mathematics, 2011.

Caldwell, Janet H., Karen Karp, and Jennifer M. Bay-Williams. *Developing Essential Understanding of Addition and Subtraction for Teaching Mathematics in Prekindergarten–Grade 2.* Essential Understanding Series. Reston, Va.: National Council of Teachers of Mathematics, 2011.

Caliandro, Christine Koller. "Children's Inventions for Multidigit Multiplication and Division." *Teaching Children Mathematics* 6 (February 2000): 420–24, 426.

Carey, Deborah A. "*The Patchwork Quilt:* A Context for Problem Solving." *Arithmetic Teacher* 40 (December 1992): 199–203.

Cooney, Thomas J., Sybilla Beckmann, and Gwendolyn M. Lloyd. *Developing Essential Understanding of Functions for Teaching Mathematics in Grades 9–12.* Essential Understanding Series. Reston, Va.: National Council of Teachers of Mathematics, 2010.

Dougherty, Barbara J., Alfinio Flores, Everett Louis, and Catherine Sophian. *Developing Essential Understanding of Number and Numeration for Teaching Mathematics in Prekindergarten–Grade 2.* Essential Understanding Series. Reston, Va.: National Council of Teachers of Mathematics, 2010.

Englert, Gail R., and Rose Sinicrope. "Making Connections with Two-Digit Multiplication." *Arithmetic Teacher* 41 (April 1994): 446–48.

Flournoy, Valerie E. *The Patchwork Quilt.* New York: Dial Books for Young Readers, 1985.

Fuson, Karen C. "Toward Computational Fluency in Multidigit Multiplication and Division." *Teaching Children Mathematics* 9 (February 2003): 300–305.

Graeber, Anna O. "Research into Practice: Misconceptions about Multiplication and Division." *Arithmetic Teacher* 40 (March 1993): 408–11.

Hendrickson, A. Dean. "Verbal Multiplication and Division Problems: Some Difficulties and Some Solutions." *Arithmetic Teacher* 33 (April 1986): 26–33.

Lannin, John, Amy B. Ellis, and Rebekah Elliott. *Developing Essential Understanding of Mathematical Reasoning for Teaching Mathematics in Prekindergarten–Grade 8*. Essential Understanding Series. Reston, Va.: National Council of Teachers of Mathematics, forthcoming.

Lobato, Joanne, and Amy B. Ellis. *Developing Essential Understanding of Ratios, Proportions, and Proportional Reasoning for Teaching Mathematics in Grades 6–8*. Essential Understanding Series. Reston, Va.: National Council of Teachers of Mathematics, 2010.

National Council of Teachers of Mathematics (NCTM). *Principles and Standards for School Mathematics*. Reston, Va.: NCTM, 2000.

——. *Curriculum Focal Points for Prekindergarten through Grade 8 Mathematics: A Quest for Coherence*. Reston, Va.: NCTM, 2006.

——. *Focus in High School Mathematics: Reasoning and Sense Making*. Reston, Va.: NCTM, 2009.

National Mathematics Advisory Panel. *Foundations for Success: The Final Report of the National Mathematics Advisory Panel*. Washington, D.C.: U.S. Department of Education, 2008.

Quintero, Ana Helvia. "Children's Conceptual Understanding of Situations Involving Multiplication." *Arithmetic Teacher* 33 (January 1986): 34–37.

Sowder, Judith T. "Estimation and Number Sense." In *Handbook of Research on Mathematics Teaching and Learning*, edited by Douglas A. Grouws, pp. 371–89. New York: Macmillan, 1992.

Thiessen, Diane. *The Wonderful World of Mathematics: A Critically Annotated List of Children's Books in Mathematics*. 2nd ed. Reston, Va.: National Council of Teachers of Mathematics, 1998.

Thiessen, Diane. *Exploring Mathematics through Literature: Articles and Lessons for Prekindergarten through Grade 8*. Reston, Va.: National Council of Teachers of Mathematics, 2004.

Weiland, Linnea. "Matching Instruction to Children's Thinking about Division." *Arithmetic Teacher* 33 (December 1985): 34–35.

# Titles in the Essential Understandings Series

The Essential Understanding Series gives teachers the deep understanding that they need to teach challenging topics in mathematics. Students encounter such topics across the pre-K–grade 12 curriculum, and teachers who understand the big ideas related to each topic can give maximum support as students develop their own understanding and make connections among important ideas.

---

*Developing Essential Understanding of–*

*Number and Numeration for Teaching Mathematics in Prekindergarten–Grade 2*
ISBN 978-0-87353-629-5     Stock No. 13492

*Addition and Subtraction for Teaching Mathematics in Prekindergarten–Grade 2*
ISBN 978-0-87353-664-6     Stock No. 13792

*Rational Numbers for Teaching Mathematics in Grades 3–5*
ISBN 978-0-87353-630-1     Stock No. 13493

*Algebraic Thinking for Teaching Mathematics in Grades 3–5*
ISBN 978-0-87353-668-4     Stock No. 13796

*Multiplication and Division for Teaching Mathematics in Grades 3–5*
ISBN 978-0-87353-667-7     Stock No. 13795

*Ratios, Proportions, and Proportional Reasoning for Teaching Mathematics in Grades 6–8*
ISBN 978-0-87353-622-6     Stock No. 13482

*Functions for Teaching Mathematics in Grades 9–12*
ISBN 978-0-87353-623-3     Stock No. 13483

*Expressions, Equations, and Functions for Teaching Mathematics in Grades 6–8*
ISBN 978-0-87353-670-7     Stock No. 13798

*Mathematical Reasoning for Teaching Mathematics in Prekindergarten–Grade 8*
ISBN 978-0-87353-623-3     Stock No. 13794

## Coming in 2012:

*Developing Essential Understanding of–*

*Geometry for Teaching Mathematics in Prekindergarten–Grade 2*

*Geometry for Teaching Mathematics in Grades 3–5*

*Geometry for Teaching Mathematics in Grades 6–8*

*Geometry for Teaching Mathematics in Grades 9–12*

*Reasoning and Proof for Teaching Mathematics in Grades 9–12*

*Data Analysis and Statistics for Teaching Mathematics in Grades 6–8*

*Statistics for Teaching Mathematics in Grades 9–12*

Visit www.nctm.org/catalog for details and ordering information.